Pedestrian Fall Safety Assessments

In-Ju Kim

Pedestrian Fall Safety Assessments

Improved Understanding on Slip Resistance Measurements and Investigations

Springer

In-Ju Kim
Department of Industrial Engineering
and Engineering Management
University of Sharjah
Sharjah
United Arab Emirates

ISBN 978-3-319-56241-4 ISBN 978-3-319-56242-1 (eBook)
DOI 10.1007/978-3-319-56242-1

Library of Congress Control Number: 2017937697

Printed on acid-free paper

This Springer imprint is published by Springer Nature
The registered company is Springer International Publishing AG
The registered company address is: Gewerbestrasse 11, 6330 Cham, Switzerland

To my parents, Young-Kyun Kim and Chang-Rae Kang, and my wife, Eun-Eun Oh, and my daughter, Sho-Young Sabrina Kim

Foreword I

Significant impacts and concerns caused by fall incidents have been globally recognized due to a large number of fatal and non-fatal injuries and the heavy burden of associated costs. There have been prolonged efforts in order to reduce the frequency and severity of fall incidents especially in advanced countries, where accidental falls are increasing year by year. Most people are liable to consider as accidental falls as unavoidable incidents and to blame themselves for being careless, when they slip or trip while walking. We, safety researchers, think that those accidental falls may not necessarily lie with pedestrians, and we are convinced that accidental falls with respect to slips, trips and falls can be reduced if we can keep on sharing multi-disciplinary efforts to prevent the fall incidents.

Dr. In-Ju Kim is an internationally recognized researcher in the area of industrial ergonomics especially for slips, trips and falls and injury prevention. After he was conferred a Ph.D. degree for the doctoral thesis entitled "A new tribological paradigm for characteristic pedestrian slip resistance properties" from the University of Sydney in Australia in 2001, he has worked for broad areas of ergonomics, human factors, applied biomechanics, and sports engineering and technology in a number of research and industry projects from Australia, the UK, the USA and Saudi Arabia over the last 15 years. I have had a professional relationship with him, and have done research in the areas of occupational safety and health, and have co-authored a number of peer-reviewed publications (2 journal papers and 5 conference papers). Over his career, he has participated in several academic disciplines, and gained much practical industrial experience.

This book mainly explains slip resistance properties from an engineering viewpoint where principal and deeper understandings on the shoe-floor friction mainly are based on his research achievements and his past experience. This book also will provide a novel understanding of the complex nature of slip resistance behaviour between shoe and floor and human interaction with slippery walking surfaces, and also a new concept to understand floor surface roughness for optimal slip resistance performance. It is clear that this book is very useful for safety researchers, safety practitioners, safety engineers, architects, building owner, shoe makers, flooring companies and university students who are interested in newly applied tribology technology to interactions between frictional surface properties and human gaits.

Hisao Nagata, Ph.D.
The Ohara Memorial Institute for Science of Labour
Former Head of Human Factors and Risk Management
Research Group, Japan National Institute of Occupational
Safety and Health (JNIOSH), Japan

Foreword II

I have met Dr. In-Ju Kim about 17 years ago at a conference held by the Liberty Mutual Research Institute related to the occupational fall prevention topic in Hopkinton MA in 2000. Since then I have followed and admired his work in "Tribology" of shoe/floor interfaces and how this research could help transform the public safety in terms of fall accidents and pedestrian safety.

This book, a representative of his knowledge domain, uncovers pertinent information related to slip induced fall accidents by elaborating on the complex nature of slip resistance properties of shoes, floors, and other elements critical to fall prevention and safety in the workplace. Use of a unique multi-factorial approach to better understand slip and fall accidents by considering the tribology and biomechanics is elaborated in the book.

I believe In-Ju's training and enthusiasm for falls research allows him to bring a high level of sophistication in each of these areas making it likely that this work with have a high impact in the field of fall prevention ultimately leading to reduction on occupational falls and pedestrian safety.

Stemming from his background in biomechanics and tribology, In-Ju has fused traditionally separate fields of tribology and fall prevention to provide a unique solution to work-related slips and falls. In-Ju has successfully identified a method to ascertain fall risks given tribological/biomechanical interactions. This is an important contribution to the field of ergonomic and biomechanics and fall prevention.

In summary, I highly encourage to read this book if you are interested in fall prevention and pedestrian safety, this book can shed light on the dark and cold world of accidental falls.

Prof. Thurmon E. Lockhart, Ph.D.
School of Biological and Health Systems Engineering
Arizona State University, Tempe, AZ, USA

Preface

Background and Motivation

Fall incidents from slips or trips have been recognized as a major threat to the safety of individuals not only in industry but also in daily living. They represent the second leading cause of accidental death, after motor vehicle accidents. Measurements of slipperiness, specifically slip resistance performance, have shown a significant role in identifying slip and fall incidents, in understanding slip and fall mechanisms, and in the development and evaluation of slip and fall prevention strategies.

One of the most generally and commonly practiced methods for the fall safety assessment is to measure a shoe-floor grip or slip resistance property as a form of coefficient of friction. Although the concept of friction is relatively simple and straightforward, solving the real-world problems on slip and fall incidents are a quite complex and challenging task. Therefore, this book aims to offer readers to uncover valuable information for a better understanding on the multifaceted nature of slip resistance properties amongst the shoes, floors and environments, learn objective ways to measuring slip resistance properties and consequently improve pedestrian fall safety assessments.

This book is intended to be an applied engineering guidebook in which the presented concepts and information on slip resistance measurements are provided with a number of graphical forms. The associated quantifying equations and formulae have kept as simple as possible. They controlled to those encountered in lecture and laboratory courses taken in the undergraduate engineering education.

This book is also written for two reader groups: (1) technical and (2) non-technical audiences. Chapter summaries at the end of each chapter are simplified reviews of the chapters' contents and important issues, which are intended for the non-technical reader.

This book brings some of the most important current research related to slips and falls and slip resistance measurements. The book can partly be a textbook and partly a monograph. It is a textbook because it gives a detailed introduction to slip resistance measurement techniques and applications. It is simultaneously a monograph because it presents several new concepts, theories, results, and further developments on slip resistance measurements and fall safety assessments. As a result of its twofold characters, this book is likely to be of interest to undergraduate and postgraduate students. This book can also be used as a handbook to engineers, industry safety consultant and practitioners, and scientists in the field of safety and industrial engineering. One can read this book through sequentially, but it is not necessary since each chapter is essentially self-contained, with as few cross-references as possible.

Main Aims of This Book

This book aims to improve the validity and reliability of slip resistance measurements from an engineering point of view where principal understandings on the shoe-floor friction and wear behaviours can be made. Therefore, this book proposes readers to find valuable information for better understanding of the complex nature of slip resistance properties amongst the shoes, floors and environments, discover objective ways to measuring slip resistance properties and learn to improve pedestrian fall safety assessments.

Readers may not only acquire solid theoretical foundations for accounting the underlying complex mechanisms of slip resistance properties but also enhance understandings on the consistency and rationality of the pedestrian fall safety measurements. The key features of this book for the readers are to

- Identify major problems of the existing methodologies for the evaluation of pedestrian slip resistance properties;
- Understand friction and wear behaviours of shoes and floors and their interactive mechanisms involved at the sliding interface between them.
- Recognize the effects of floor surface finishes on slip resistance properties and determine design ideas with operational levels of floor surface roughness for optimal slip resistance performance under a range of slippery environments.

Construction of This Book

This book begins with a discussion on slip resistance measurements as a format of the coefficient of friction (COF), the most widely used definition and classification for fall safety assessment and their tribo-physical characteristics and consequences on slip resistance performance. Furthermore, this book provides in-depth overviews on friction mechanisms and suggests practical design recommendations for pedestrian walkway surfaces to prevent falls and fall-related incidence. The following areas are covered in each chapter:

Chapter 1
There have been sustained efforts to understand the main causes of slip and fall incidence in order to reduce their injuries and severities. It has been found that slip resistance performance between the footwear and underfoot surface is of great importance for preventing fall incidence and has been measured as a form of a coefficient of friction (COF). In this milieu, knowledge about friction demand and friction available has been recognized as the key factor in slip evaluation. Since the COF measurements at the sliding interface have been adopted to determine whether a slip is to occur, there has been ambiguity in the interpretation of COF measurement results. It has been found that any slip resistance measurement results have unique characteristics to a specific combination of the shoe-floor-contamination tested and constant changes during the tests. Hence, observing slip resistance properties with a simple friction measurements has obvious difficulty as an indicator for identifying fall hazards between the footwear and underfoot surfaces.

This chapter demonstrates that future research for the pedestrian fall prevention requires improving the validity and reliability of slip resistance measurements from an engineering point of view where principal understandings on the shoe-floor friction and wear behaviours can be made. This chapter also discusses that comprehensive investigations for the surface analyses of the shoes and floors with dynamic friction measurements are required to understand mechanical and physical behaviours of the shoe-floor friction and wear systems.

Chapter 2
Slip resistance property between the footwear and underfoot surface is of great importance for assessing slip and fall incidence and has been measured as a form of a coefficient of friction (COF). Hence, knowledge about friction demand and friction available has been recognized as a key factor to fall safety measures. Since the COF measurements at the sliding interface between the floor surface and shoe heel have been adopted to determine whether a slip is to occur, there has been ambiguity in the interpretation of the COF readings. The recent studies have found that any slip resistance measurements have (1) characteristics peculiar to a specific combination of the shoe-floor-environment measured and their interaction at the sliding interface, and (2) constantly varied during their service times. Hence, there is an inherent risk in relying upon a single COF result to provide an indication of the slip resistance properties between the shoe and floor surface.

In this sense, this chapter is focused on improving the validity and reliability of slip resistance measurements. To achieve this goal, the problem of slip resistance analysis has been approached by a tribo-physical point of view where a principal understanding on the floor-shoe friction mechanism can be made. This chapter also covers fundamental aspects of slip resistance properties and surface analyses for the floors and shoes with their mechanical and physical characteristics.

Chapter 3
As clearly deliberated in Chaps. 1 and 2, simple friction measurements could misrepresent accurate slip resistance properties between the shoe and floor. Facilitated routine friction measurements from laboratory environments could also oversimplify the intrinsic characteristics of slip resistance. Although there has been considerable progress in understanding the fundamental features of frictional behaviours of the shoes and floors, it is probably true to say that none of the COF measurements reported to date can be regarded as final objective values for a given shoe-floor-contaminant combination.

As long as the controversy around friction measurement as a format of COF remains, improvements in the principal concepts and methodologies on slip resistance measurements are urgently required. Therefore, this chapter principally focused on broadening the knowledge base and developing new ideas on which improvements in the validity and reliability of slip resistance measurements may be made.

Chapter 4
An adequate level of traction or slip resistance at the shoe-floor sliding interface is required for unperturbed ambulation. Without the presence of friction, ambulation could simply not occur. The classical model of friction defines slip resistance behaviours within simple parameters and is limited in its ability to explain the mechanics present at the shoe-floor sliding interface. In the context of human ambulation, however, friction is a complex phenomenon and contains multi-factorial mechanisms. Over time periods, the classical model of friction has developed into a paradigm which accounts for both human and environmental friction components; the relationship between them determines the propensity for a slip to occur.

However, as demonstrated above, surface topographies of both shoe and floor can be largely changed by friction-induced wear developments. As a result, this would significantly affect slip resistance properties. Therefore, surface finishes of the shoe and floor should be monitored routinely to maintain and provide the best slip resistance performance against specific walking/working environments.

This information may provide more reliable results to manage pedestrian fall safety than measuring slip resistance alone. Such approaches highlight the need for developing enhanced concepts and methods for reliably characterizing slip resistance properties. This should be based on thorough understanding of the complex nature of frictional behaviours between the shoe and floor surface, their related tribo-physical characteristics, and their interactive impacts on slip resistance performance.

Chapter 5
This chapter discusses the factors that affect slip resistance properties between the shoe heel and the floor surface. The factors are mainly classified into two components: adhesion and deformation in terms of frictional force. The contribution of the resultant frictional force totally depends on the various elements such as normal pressure, sliding force and surface texture. Amongst all these elements, however, the surface characteristics and their interactions are eventually major reasons to affect frictional and wear events of the sliding interface between the shoe outsole and floor surface. Because the surfaces of almost all solids are far from smooth under the microscopic view, even the surfaces of shoes and pedestrian floorings seem to have coarser and rougher surface features. It is, therefore, necessary to investigate the variations of surface characteristics with their friction and wear behaviours.

The above descriptions of the possible forms of friction and wear behaviours clearly demonstrated why it might never be possible to predict accurately their tribo-physical characteristics between two solid materials, particularly in the case of sliding friction between the shoe heel and floor surface. A systematic approach was discussed in which friction and wear processes between the shoe heel and floor surface during the prolonged sliding events were broken down into several elemental courses forming a feedback loop and a newly developed model for the wear mechanisms between the shoe and floor surface was suggested.

It is considered that theoretical developments for the shoe-floor contact surfaces and their tribo-physical characteristics have not yet reached a mature stage, where it would be possible quantitatively to predict friction and wear behaviours from known surface characteristics; but this would be a useful diagnostic tool and be a step forward to identify the complex issues. In this chapter, therefore, the origin of friction forces was examined and related tribo-physical features such as wear evolutions at the sliding interface between the shoe and floor surface were extensively explored. However, the discussion was mainly focused on understanding for the friction and wear behaviours of unlubricated solids in sliding motions as a first instance.

Chapter 6
The measurement and interpretation of slip resistance properties should be based on full understanding of the relevant tribo-physical characteristics and involved mechanisms as an essential prerequisite because the friction measurement amongst the shoe, floor and environment are not a simple matter. In this sense, it would be a constructive attempt to study the topographic features of surfaces, their contact-sliding mechanisms and related tribo-physical behaviours. In order to recognize tribological processes involved at the sliding interface between the shoe and the floor surface, this chapter discusses how the two surfaces interact when they are loaded together. Surface analyses and relevant background information were comprehensively reviewed with different measuring instruments to quantify topographic aspects of the shoe and floor surfaces.

Several geometrical models were comprehensively reviewed and related theoretical issues were also carefully discussed. The concept on changes in surface topographies of two interacting bodies as a result of sliding friction actions was applied to the sliding interface between the shoe and floor. Based on the main assumptions with thorough knowledge foundations for the contact-sliding mechanisms between two surfaces, a theory model on the contact-sliding mechanism between the shoe heel and the floor surface was suggested. A perfect contact-sliding model to predict slip resistance performance between the shoe and the floor surface has not been developed yet, but tribological approaches seem to be a worthwhile attempt to overcome limitations on the existing researches and practices for the measurements of pedestrian fall safety.

Chapter 7

Increasing traction properties of the floor surface would be desirable as a general rule, but a very high level of slip resistance may impede safe and comfortable ambulation. There is a lack of evidence whether traction properties are linearly correlated with surface features of the floor or what levels of floor surface finishes are required to effective control of slipperiness. It is also scarce to find studies and/or guidelines on the operational ranges of floor surface roughness required for optimal slip resistance performance.

The main objectives of this chapter are to investigate the effects of floor surface finishes on slip resistance performance under different environmental and shoe-type conditions and identify operational ranges of floor surface roughness as practical design information for the effective control of fall incidents. A theory model of three operative zones was suggested to characterize functional levels of floor surface roughness on slip resistance performance. To test the theory model, dynamic friction tests were conducted using 3 shoes and 9 floor specimens under 4 different environments: clean and dry, wet, soapy and oily conditions. The test results showed that significant effects of floor-type on DFCs were found in the polluted environments. As compared to the floor-type effect, the shoe-type effect was relatively small. Slip resistance performance was significantly affected by and well correlated with the floor surface roughness under the soapy and oily environments. Polynomial regression analyses amongst the floor surface roughness, DFCs and environments allowed to estimating operational ranges for optimal slip resistance performance.

Floor surfaces with around 17 μm to 52 μm and 35 μm to 52 μm in $\boldsymbol{R_a}$ roughness parameter most likely represented the lower and upper bounds of operational ranges for the best slip resistance managements under the soapy and oily surface conditions, respectively. The test result also identified that the oily environment required twice rougher floor surface than the soapy one in their lower boundary roughness scales: 35 μm vs. 17 μm in $\boldsymbol{R_a}$ roughness parameter. On the other hand, the upper bound of floor surface roughness showed the same ranges of surface roughness scales: 52 μm in $\boldsymbol{R_a}$, 300 μm in $\boldsymbol{R_t}$, and 180 μm in $\boldsymbol{R_{tm}}$ roughness parameters under the three lubricated environments. However, there was a lack of correlations between the surface roughness and slip resistance properties under the clean and dry and wet surface conditions.

The inclusive results from this chapter also clearly display that the proposed concept on the operational ranges for the floor surface roughness may have practical design implications for floors and floor coverings to reduce slip and fall hazards.

Chapter 8

This chapter suggests recommendations for the future studies of pedestrian fall incidents and their investigations, measurements, interpretations and prevention strategies. Specifically, this chapter suggests that physical accuracy and validity of the developed shoe-floor-environment sliding friction model need to be improved and expanded to include a diverse range of shoes, floor types and environmental conditions. The chapter proposes exclusive ideas on how the future generation of tribo-physical model(s) for the shoe-floor-environment can be further improved to become a useful tool(s) for the prevention of slip and fall incidence.

Final Remarks

As a researcher in this field, I am honoured to be writing a book with such a fascinating and exciting topic. I would like to thank the reviewers, who have committed so much towards the publication of this work. Without their invaluable reviews, this book could not have been written.

Special thanks go to Dr. Dieter Merkle, Vice President of the Applied Sciences Division from Springer Germany and Mr. Anthony Doyle from Springer UK for publishing my book. I also would like to thank Ms. Padmavathi Jayajeevan for her kind assistance in producing this book.

Sharjah, United Arab Emirates In-Ju Kim Ph.D.

Contents

Glossary Terms, Abbreviations and Acronyms

ACA	Apparent geometric area of contact
ACOF	Available COF
AFM	Atomic force microscopy
ANSI	American National Standards Institute
ANOVA	Analysis of variance
AS/NZS	Australian/New Zealand Standards
BRS	Building Research Station
BSI	British Standards Institution
CLA	Centre line average, R_a
COF	Coefficient of friction
COS	Committee of Standards
CTIOA	Ceramic Tile Institute of America
DCOF	Dynamic (or kinetic) coefficient of friction
DFC	Dynamic friction coeff*f*icient
EHL	Elastohydrodynamic lubrication
FP	Force plate
GLC	Greater London Council
H	Hardness
HF	Frictional (or horizontal) component of the resultant force
HPS	Horizontal Pull Slipmeter
HSC	High spot count
HSE	Health and Safety Executive
ISO	International Organization for Standardization
LSCM	Laser scanning confocal microscope
NBS	National Bureau of Standards
NCA	Nominal contact area
NT	Number of test
OECD	Organization for Economic Cooperation and Development
OSHA	Occupational Safety and Health Administration
PIAST	Portable inclinable articulated strut slip tester

PSRT	Programmable slip resistance tester
PSS	Portable slip simulator
PTSRVs	Pendulum test slip resistance values
PTV	Pendulum test value
PU	Microcellular polyurethane
RCA	Real area of contact
RCOF	Required COF
RLIM	Reflected light interference microscopy
RMS	Root mean square, $\boldsymbol{R_q}$
SCOF	Static coefficient of friction
SEM	Scanning electron microscope
SFC	Static coefficient of friction
S/N	Signal-to-noise ratio
SAS	Statistical analysis system
SSR	Sustainable slip resistance
STM	Scanning tunnelling microscopy
TCNA	Tile Council of North America
TRCA	Total real area of contact
TRRL	Transport and Road Research Laboratory
UCOF	Utilized coefficient of friction
UFTM	Universal Friction Testing Machine
UKSRG	UK Slip Resistance Group
VF	Vertical component of the resultant force
VIT	Variable incidence tribometer
a_i	Radius of each circular contact spot
A	Real area of contact
A_{ai}	Projection of an enclosed surface that is a real contact area
A_{act}	Summation of individual areas at the summits of asperities
A_{app}	Apparent area
A'	Individual contact area
Ai	Discrete area
An	Sum of each of the discrete areas Ai
Da	Density of contour
d	Separation between the reference plane and the flat surface
E	Young's moduli
F	Tangential (or friction) force
F_A	Adhesion term in a frictional force
F_{adh}	Adhesion force
F_D	Deformation term in a frictional force
F_{def}	Deformation force
F_N	Normal force
F_S	Sliding traction force
I	Mean sum of profile peaks
k	Critical shear stress of this material
K	A constant

L	Nominal length of a surface
L'	Actual length of a surface
n	Number of contacts
P	Mean actual pressure
P_{app}	Apparent or nominal pressure
P_o	Yield pressure of the softer one of the two materials
py	Yield pressure
$P(y)$	Amplitude density function
q	Constant of proportionality
r	Asperity radius of curvature
R_a	Centre line average roughness parameter
R_k	Kurtosis roughness parameter
R_{max}	Maximum peak-to-valley height within a sampling length
R_p	Maximum departure of the profile above the mean line
R_{pm}	Mean of maximum departure of the profile above the mean line
R_q	Root measure square roughness parameter
R_{sk}	Skewness roughness parameter
R_t	Maximum peak-to-valley height roughness parameter
R_{tm}	Maximum mean peak-to-valley height roughness parameter
R_v	Maximum departure of the profile below the mean line roughness parameter
R_{vm}	Mean of maximum departure of the profile below the mean line roughness parameter
s	Friction force per unit area (shear strength)
$tan\ \alpha$	Ratio of energy dissipated to energy stored per cycle
V	Normal (or vertical) component of the resultant force
W	Normal load
W_i'	Individual load
Y	Yield stress
z	Height of an individual asperity above the reference plane
z_L	Asperity lower than a limiting height
α	Rigid conical asperity of semi-angle
Δ_a	Average slope of a surface profile
Δ_q	RMS slope of a surface profile
θ	Asperity attack angle
μ	Coefficient of friction
λ	Mean wavelength
λa	Average wavelength
λq	Root mean square (RMS) wavelength

List of Figures

List of Tables

Chapter 1
Introduction

1.1 Backgrounds

Pedestrian fall incidence resulting from slips or trips is one of the foremost causes of fatal and non-fatal injuries that take more loss of functionality. It occurs at any age group including healthy people. Fall injuries are also one of the major outcomes in surveys on serious occupational incidents as well as one of the most common geriatric syndromes threatening the independence of older people (Dias et al. 2011; Perez-Jara et al. 2012; Demura et al. 2013; Whitney et al. 2013). More than one-third of adults 65 and older fall each year in the United States (Hornbrook et al. 1994; Hausdorff et al. 2001).

According to the U.S. Bureau of Labour Statistics, 666 workers lost their lives due to fatal falls in 2011 (Maurer 2012). Falls are regarded as the secondleading cause of accidental deaths worldwide and are a major cause of personal injuries, especially for the elderly (Ozanne-Smith et al. 2008; WHO 2012). The annual direct cost of injuries related to slips and falls in the working environment is about $5.7 billion (Liberty Mutual 2003; DiDomenico et al. 2007). Additionally, lost times due to falls negatively affect productivities and business (WSDLI 2010). Hence, employers in the industries and workplaces are in great need of solutions that overcome or prevent the unintentional injuries due to slips and falls.

Health and Safety Executive (HSE) in the UK stated that 95% of the major results from slip and fall incidents was broken bones and up to one in three of the major work-related accidents was the result of slips and/or trips in the workplaces (Mills et al. 2009; HSE 2013). In Australia, slip and fall incidents are also a leading cause of work-related injuries (Safe Work Australia 2013). However, the concern is not only related to the high incidence of falls in older people but rather the combination of high incidence and susceptibility to injury (Rubenstein and Josephson 2006; Peel et al. 2008; Julian Robinson 2016). The slip and fall incidence also results in thousands of injuries each year with the most common injuries being musculoskeletal, cuts, bruises, fractures, and dislocations (Safe Work Australia 2012).

I.-J. Kim, *Pedestrian Fall Safety Assessments*,
DOI 10.1007/978-3-319-56242-1_1

Figure 1.1 shows a photograph that Pope Francis missed a step and fell to the ground as he was coming to an open-air altar to celebrate Mass at Poland's holiest shrine of Jasna Gora on 28th July 2016.

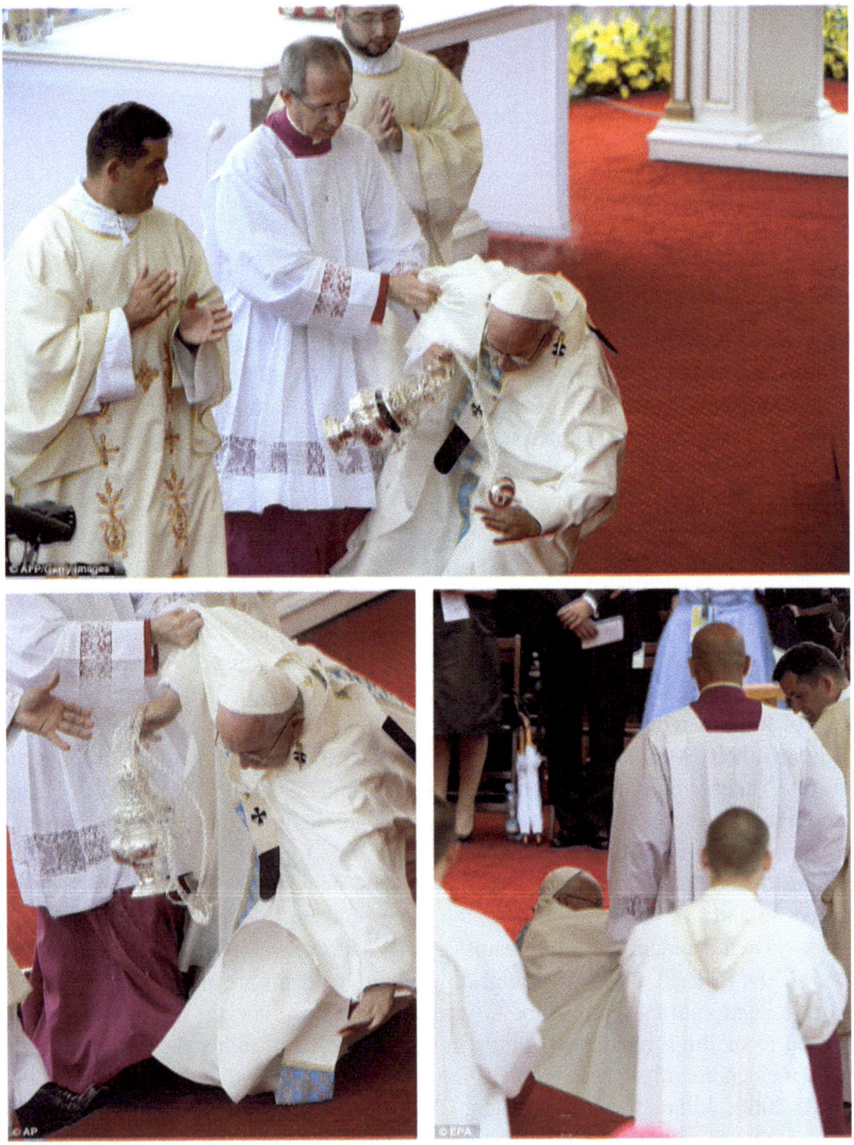

Fig. 1.1 Pope Francis takes a fall at the beginning of the Holy Mass in the Shrine of Czestochowa on the occasion of the 1050th anniversary of the baptism of Poland 28 July 2016 (Julian Robinson 2016)

The importance of the fall incident has been globally recognised due to the magnitude of problems and associated costs. There have been prolonged efforts to identify and understand the main causes of such incidence in order to reduce their frequency and severity throughout the world (Redfern et al. 2001). It has been found that one of the most common precipitating events leading to a fall is a loss of traction or slip resistance between the shoe sole/heel and floor surface.

As a result, a shoe-floor grip or slip resistance property has been commonly measured as a form of coefficient of friction (COF) (Cohen and Compton 1982; Harris and Shaw 1988; Proctor and Coleman 1988; Chaffin et al. 1992; Kim and Smith 2000, 2003; Kim et al. 2001, 2013; Kim 2002, 2003a, b, 2006a, b, 2015, 2016; Kim and Nagata 2008). Hence, information on required friction and accessible friction has been recognised as a key element for the fall safety assessment.

Since the COF measurements between the shoe sole/heel and floor surface were commonly adopted to determine whether a slip was to occur, there have been uncertainties in the interpretation of friction measurement results (Kim 2006b; Kim and Nagata 2008). Importantly, its analyses, measurements, and interpretations have been generally misguided in most research and practice works for the assessments of fall safety. That is, COF results from any slip resistance measurement show:

(1) unique characteristics of a specific combination amongst the shoe, floor, and environment; and
(2) constant changes during the entire measurements.

Although the concept of friction is relatively simple and straightforward, its measurement, analysis, and clarification for the solutions of real-world problems on the pedestrian fall incidence are quite challenging tasks. However, one of the most important aspects to address is that the COF index or quantity seems not a good indicator to detect slip resistance properties because it becomes fundamentally noisy and continuously changes as a function of a complex array of tribo-physical behaviours between the shoe and floor (Kim and Smith 2000, 2003). In addition, frictional phenomena observed at the sliding interface between the shoe heel and floor surface involve multiple characteristics and combine various sub-mechanisms (Kim and Smith 2000, 2003; Kim 2004a). Therefore, there is an inherent risk to rely upon a single COF threshold to provide an indication of the fall safety.

Those concerns and issues on slip resistance measurements for the pedestrian fall safety assessments have led to finding a new insight to enhance our understanding of the multi-dimensional properties of slip resistance. The recent research suggests that a tribological classification may provide an objective alternative to overcome the current problems on slip resistance evaluations. Therefore, this book robustly discusses limitations on the present concept for slip resistance measurements and analyse the seriousness of misinterpretations on slip resistance properties that are mainly triggered by oversimplified perceptions on friction behaviours between the shoe heels and floor surfaces. Based on such critical reviews, this book proposes a new paradigm for future research on the slip resistance measurements.

1.2 Major Issues on Slip Resistance Measurements

It has been found that one of the most common precipitating events leading to a fall is a loss of traction or slip resistance between the shoe sole/heel and the floor surface, followed by trips, misstep, loss of support, and postural overextension (Cohen and Compton 1982). Hence, a grip or slip resistance property between the shoe and floor has been commonly measured as a form of COF (Cohen and Compton 1982; Harris and Shaw 1988; Proctor and Coleman 1988; Chaffin et al. 1992; Kim and Smith 2000, 2003; Kim et al. 2001, 2013; Kim 2006a, b, 2015, 2016; Kim and Nagata 2008). Accordingly, information on required friction and accessible friction has been identified as a crucial element so that knowledge about the friction demand and the friction available has been recognised as a major key factor for slip safety estimation.

From the time when this concept was adopted for assessing the pedestrian slip safety, a number of friction measuring apparatus and/or devices have been found in the literature, and some of them are commercially available. To this time, however, none of them is internationally adopted as a standard model and/or a tester because each of them has some advantages and disadvantages in their designs and operating functions. The versatility of different devices for slip resistance measurements has been compared and commented by several studies (National Research Council 1961; Bring 1964; Brungraber 1976; Braun and Brungraber 1978; Bring 1982; Cohen and Compton 1982; Andres and Chaffin 1985; English 1990; Martin 1992).

As mentioned above, the COF index is not a constant quantity because initial surface features and topographic characteristics of both shoes and floors are frequently and significantly modified from the first moment of contact by repetitive friction and wear developments. As a result, frictional properties become noisy and continuously change as a function of a complex array of tribo-physical phenomena amongst the shoes, floors, and environments (Kim and Smith 2000, 2003). Thus, there have been large disagreements in the interpretation of measured results of slip resistance properties. Importantly, its measurements, evaluations, and clarifications have been misguided in many research and practices for fall safety assessments.

Friction is a direct and clear concept but involves complicated mechanisms. From a historic point of view, it was considered that slip resistance measurements between the shoe and floor would at first be a simple matter. However, it has been realised that the development of a standard (national and international) on slip resistance measurements is an intricate matter and needs to consider a wide range of factors comprised at the sliding friction interface. The following factors seem to be some of the upmost important issues to consider:

(1) People walk differently and have different COF requirements.
(2) Personal safety is related to one's perception and awareness on slip and fall risksfall risk assessmentassessment.
(3) Any slip resistance measurement has unique characteristics to a specific shoe-floor-environment combination.

As demonstrated in the above, therefore, it becomes clear that there is an obvious difficulty to use a single friction index as a safety indicator for the slip resistance performance amongst the shoe, floor, and environment. However, most research studies and industry practices for the pedestrian fall safety assessment still make the simple demand that there should be a minimum COF threshold of at least 0.4 or 0.5 available between the shoe and floor surface. Therefore, this book focuses on

(1) understanding fundamental aspects of slip resistance properties between the shoe and floor,
(2) filling the knowledge gaps to overcome the current limitations and mistakes on slip resistance measurements, and
(3) suggesting a novel concept for engineering/technical solutions on the slip resistance measurements and analyses.

To achieve the above aims, this book deals with a large volume of information to discuss tribo-physical characteristics such as friction and wear behaviours of the shoes and floors with surface analyses. This information applies to recognise multifaceted characteristics of sliding friction and wear mechanisms, their impacts on slip resistance measurements, and consequences on slip resistance performance between the shoe and floor. The inclusive discussions and concept developments are sequenced as follow:

(1) Classical principles and models of tribology are used to explore friction and wear behaviours and related tribo-physical mechanisms between the shoe and floor surface.
(2) Fresh theory concepts for the analysis of surface interactions between both bodies arising from the tribological models are proposed to monitor surface changes of the shoes and floors during dynamic friction measurements and their impacts on slip resistance performance.
(3) The proposed theory notions and models are also accompanied by observing topographic changes with measuring surface roughness parameters within the contact areas of shoe heels and floor surfaces, as well as the interfacing surfaces between them.

Based on the above works, a final chapter of this book suggests a new design concept on operative ranges of floor surface roughness for optimal slip resistance controls under different risk levels of walking environments. This information may have conceivable attention for the design enhancements of floors and walkways to prevent pedestrian fall incidents.

It can be anticipated that collected information on operative ranges of floor surface roughness under diverse walking environments will be served as a reference to improve designs for the floor surface finishes and accordingly a valuable source to develop practical design information and guidelines for floor surfaces required to prevent pedestrian slip and fall incidents.

1.3 Surface Finishes Versus Slip Resistance Performance

Since 1988, many studies in the literature have emphasized the importance of surface roughness on slip resistance properties and their effects on slip resistance performance (Grönqvist 1995; Rowland et al. 1996; Kim 1996a, b, c; Chang 1998,1999; Kim and Smith 1998, 1999; Manning et al. 1998; Barry and Milburn 1999). A number of surface roughness parameters were introduced to analyse surface features of shoes and floors and measured to identify correlations between the surface texture and slip resistance properties (Kim and Smith 2000, 2003; Chang 2001, 2002; Kim et al. 2001, 2013; Kim 2004a, b, 2006a, b, 2015, 2016; Kim and Nagata 2008). Those studies stated that surface roughness of the shoes and floors significantly affected slip resistance performance under a range of walking environments.

Surface roughness offers drainage spaces to avoid squeeze film formations under polluted environments. For example, tread patterns on the heel surface can improve traction properties by providing void spaces for removing lubricants and leading to an increase in direct contact with the floor surface (Kim et al. 2013). Therefore, macro-roughness or tread patterns are commonly designed into the shoe heel and sole areas, but they become ineffective quickly after being worn (Kim et al. 2013; Kim 2015, 2016). However, floor surfaces seem to provide better slip resistance effects than shoe ones because surface roughness of the floor may offer sharper, taller, and tougher asperities in their surface finishes than shoe ones (Kim et al. 2001, 2013; Kim 2004a, b, 2015, 2016).

Although intensifying slip resistance properties of the floor surface would be desirable as a general rule, a very high level of traction or slip resistance may impede safe and comfortable ambulation (Chaffin et al. 1992). Moreover, maintaining and/or increasing the surface roughness of floors and floor coverings require high processing costs (Kim et al. 2013; Kim 2015).

However, most of the studies in the literature are still limited their main analyses to measuring surface roughness either of the shoes or the floors. As a result, geometrical interfacings and interactions between the shoe and floor surfaces were not considered to its full extent. This aspect should be fully explored to identifying their interactive natures between two coupling surfaces during sliding friction events because they are directly related to slip resistance performance.

1.4 Wear Development Versus Slip Resistance Performance

Friction-induced wear developments observed at the sliding interface between the footwear and underfoot surfaces seem to show an equally important role to slip resistance properties. Wear progress on the shoe heel and floor surface is likely to start immediately with friction measurements. The wear events largely change the surface conditions of both bodies and accordingly slip resistance performance.

For example, wear features and developments on the shoe surfaces have been quantitatively and qualitatively examined before and after the tests (Kim and Smith 2003; Kim et al. 2001; Kim 2005, 2015, 2016). Test results showed that the initially unique micro- and macro-tread patterns experienced massive changes and severe damages. The worn surfaces of shoe heels acquired dissimilar wear shapes, sizes, and patterns. The main differences in their wear developments were strongly related to the material characteristics. Findings from those studies provided a new insight concerning the primary features of shoe wear such as abrasion patterns, crack formations, ruptures, structures, and damage propagation (Kim 2015, 2016). The abrasion patterns of shoe heel/sole surfaces resulted from crack propagation at the root of the wear tongue and subsequent tearing of those tongues when they reached their maximum sizes (Kim 2016). Wear behaviours of the shoe surfaces were significantly affected not only by the rate of crack propagation along a low angle of asperity slope but also by the rate of crack propagation.

Changes in the surface features of floors with slip resistance measurements have been reported in the recent literature. Extended wear developments on smooth floor surfaces could cause buffing effects and considerable drops on COF quantities (Manning et al. 1998; Kim and Smith 2000, 2003; Kim et al. 2001, 2013). Those studies showed strong relationships between the floor surface roughness and slip resistance properties (Kim and Smith 1998, 2000, 2003; Kim et al. 2001, 2013; Kim 2015, 2016).

Derler et al. (2008) investigated the shift of COFs against various floor surfaces over a period of 30 months, in order to study short- and long-term effects of use and maintenance. They reported that mechanical abrasions and coatings by care products led to continuous reductions of slip resistance properties, which were typical outcomes for many floor surfaces in use. Their complex interactions against shoe surfaces led to considerable local variations of the surface topographies of floors due to wear growths.

As a result, surface finishes of the floors and walkways were largely modified from their initial ones (Leclercq and Saulnier 2002; Kim and Nagata 2008; Kim et al. 2013). These results were also confirmed by other studies that included field measurements with a range of floors and floor coverings from different test sites, mechanical wear, soiling, and maintenance (Kim and Smith 2000, 2003; Chang et al. 2003; Kim 2004a, 2015, 2016; Li et al. 2004; Kim and Nagata 2008; Kim et al. 2013).

Therefore, it becomes clear that wear advancements of the shoe and floor surfaces are inescapable and can substantially affect slip resistance properties. Surprisingly, however, there are almost no studies on how the footwear and walkway surfaces are influenced by friction-induced wear developments during slip resistance measurements. Despite the significance of this issue, its fundamental perception on wear behaviours, associated tribo-physical characteristics, and their impacts on slip resistance performance have remained as an unexplored area.

1.5 Optimal Floor Surface Finishes

Underfoot surfaces and walkways should be built to provide safe and comfortable ambulation. They also should deliver optimal slip resistance qualities against any slippery environment throughout their lifetimes. Whilst supporting and controlling slip resistance properties of the floor surfaces would be generally desirable, a specific problem may arise in the real world' walking situations. That is, with repeated walking, surface finishes of floors and walkways seem to experience large changes due to ageing of flooring materials, wear and tear, soiling, and maintenance (Kim and Smith 2000; Leclercq and Saulnier 2002). As a result, the slip resistance functions of floors and floor coverings deteriorate over time periods. Hence, increasing slip-resistance properties of the floor surface seem to be an ideal practice, but a very high level of traction or slip-resistance may impede safe and comfortable ambulation (Chaffin et al. 1992). Moreover, maintaining and increasing the surface roughness of floors and floor coverings may require high sustaining and processing costs.

Although numerous experimental and analytical studies on the prevention of slip and falls incidents are found in the literature, no theoretical concepts or models are developed to predict the effect of floor surface finishes on slip-resistance performance. In particular, it is hard to find any definitive study and design information for operational ranges of floor surface finishes required for optimal slip-resistance performance. There are also no internationally accepted guidelines and design data for operational levels of floor surface coarseness to effectively control slip resistance functioning. Therefore, it is necessary to develop a method, which can provide practical design information for the floor surface finishes against a range of walking environments.

It can be expected that collected information on operative ranges of floor surface roughness under diverse walking environments seems to be served as a reference to improve floor surface finishes and accordingly a valuable source to develop practical design information and guidelines for floor surface finishes required to prevent pedestrian slip and fall incidents.

1.6 Major Significances and Contributions

As clearly pointed out in the above discussions, the currently practised assessments for the pedestrian fall safety are mainly based on friction measurements amongst the shoes, floors, and environments. However, such approaches reveal serious limitations to accurately measure slip resistance performance.

Hence, this book focuses on broadening the knowledge base and developing novel concepts for which improvements in the validity and reliability of slip resistance measurements can be made. To achieve this goal, the existing problems on pedestrian fall safety measurements are critically assessed and discussed from a tribological point of view that may provide an objective alternative way to measuring slip resistance properties. This approach aims to cover principle understandings on the engaged tribo-physical characteristics such as friction, wear, and lubrication behaviours and mechanisms amongst the shoe, floor, and environment.

This attempt also purposes to identify how interactions of the asperities in sliding contacts can affect the surface conditions of both shoes and floors and control friction and wear behaviours amongst the shoe, floor, and environment during slip resistance measurements. This work incorporates extensive investigations of topographic characteristics of the shoe and floor surfaces and their interactive impacts on slip resistance performance. Based on such integrate efforts, this book suggests fresh theoretical concepts and models including numerical formulations for analysing the slip resistance properties.

With the above goals, some of the most important contributions that this book can offer readers are to

(1) uncover valuable information for a better understanding of the complex nature of slip resistance properties amongst the shoes, floors, and environments,
(2) learn objective ways to measuring slip resistance properties, and
(3) consequently, improve pedestrian fall safety assessments.

Therefore, it is wished that this book can not only deliver sound theoretical foundations for accounting the underlying complex mechanisms of slip resistance properties amongst the shoes, floors, and environments but also enhance the consistency and rationality of the pedestrian fall safety measurements.

1.7 Specific Aims

This book focuses on measuring, analysing and interpreting slip resistance properties from an engineering viewpoint where principal understandings on the shoe-floor friction and wear behaviours can be made. This book also includes comprehensive investigations on the surface finishes of shoes and floors with dynamic friction measurements and covers to comprehend mechanical and physical behaviours of the shoe-floor tribological system. Finally, this book suggests a new design concept to identifying operational levels of floor surface roughness for optimal slip resistance performance under a range of slippery environments.

This book also expects to achieve education and research improvements on slip safety assessments so that covers a large volume of information to discuss the subject matter on slip resistance measurements. However, there is much more to learn about the material herein. The specified aims of this book are to help readers to

(1) identify major problems of the existing methodologies for the evaluation of pedestrian slip resistance properties;

(2) characterise and analyse slip resistance properties between the shoe and floor surface from a tribo-physical point of view. This is primarily concerned with the understanding of friction and wear behaviours of shoes and floors and their interactive mechanisms involved at the sliding interface between them.
(3) assess topographic characteristics of the shoe and floor surfaces during dynamic friction measurements and their effects on the slip resistance performance.
(4) investigate geometric interactions between the shoe heel and floor surface during repetitive sliding processes.
(5) develop new concepts to analyse slip resistance properties, which can be more logical and reliable than a simple friction measurement.
(6) investigate the effects of floor surface finishes on slip resistance properties and identify operational levels of floor surface roughness for optimal slip-resistance performance under a range of slippery environments.

1.8 Limitations

When testing and analysing slip resistance performance between the shoes and floors, there are abundant choices of shoes and flooring materials available from the market. Hence, selecting the best shoes and floors for a specific profession or an industry type seems to be one of the most challenging tasks to ascertain their safety, performance, and durability issues for both products over periods.

In this sense, experimental works from this book have limited to test a small selection of footwear and floor specimens with a controlled range of surface roughness scales. Future research on the slip resistance measurements requires to testing shoes and floors with different types of materials and ranges of topographic features. This will help to systematically determine their specific design qualities and effects on slip resistance performance.

The experimental designs from this book also have involved restricted environmental conditions such as clean and dry, water-covered wet, soapsuds-covered soapy and machine oil-covered oily situations. This may limit the applicability of findings only to these types of environmental conditions. Other kinds of surface pollutants with different compositions and viscosities may result in different slip resistance performance and require changed operational levels of surface roughness.

1.9 Summary

Pedestrian fall incidence from slips or trips is a major concern. They are a primary cause of workplace injuries as well as a leading cause of injury-related deaths for the elderly age 75 and over. There have thus been prolonged efforts to understand the main causes of such incidence in order to reduce their injuries and severities.

It has been found that the most common precipitating event leading to a fall is a loss of traction between the shoe sole/heel and floor surface, followed by trips, missteps, loss of supports, and postural overextensions (Cohen and Compton 1982). Slip resistance between the footwear and underfoot surface is of great importance for assessing fall incidence and has been measured as a form of a coefficient of friction (COF). In this milieu, knowledge about friction demand and friction available has been recognised as the main key factor to slip safety estimation.

Despite many years of investigations and fabrication of numerous testing devices and tools throughout the world, there are still no internationally accepted standards for the measurements of slip resistance performance between the footwear and underfoot surfaces. Since the COF measurements at the sliding interface have been adapted to determine whether a slip is to occur, there have been large uncertainties to interpret COF results. It has been found that COF outcomes from any slip resistance measurement show:

(1) unique characteristics of a specific combination of the shoe-floor-environment; and
(2) constant changes during the whole test.

In addition, wear developments of both shoe and floor surfaces are severer than assessed and its effects on slip resistance performance are quite significant (Kim and Smith 1998). Accordingly, it becomes evident that a simple friction measurement is not a proper way to assess pedestrian fall safety. That is, friction and wear behaviours of the shoe and floor and their interactive tribo-physical characteristics should be thoroughly investigated from a perspective of fundamental causes and effects.

This book aims to improve the validity and reliability of slip resistance measurements from an engineering point of view where principal understandings on the shoe-floor friction and wear behaviours can be made. This book also includes comprehensive investigations for the surface analyses of the shoes and floors with dynamic friction measurements and covers to understand mechanical and physical behaviours of the shoe-floor friction and wear systems. As a final point, this book suggests a new design concept to identifying operational levels of floor surface roughness for optimal slip resistance performance under a range of slippery environments.

Therefore, this book attempts to not only deliver sound theoretical foundations for accounting the underlying complex mechanisms of slip resistance properties amongst the shoes, floors, and environments but also enhance the consistency and rationality of the pedestrian fall safety measurements. Through this book, the author wishes the readers to uncover valuable information for better understanding of the multifaceted nature of slip resistance properties amongst the shoes, floors, and environments, learn objective ways to measuring slip resistance properties and consequently improve pedestrian fall safety assessments.

References

Andres, R. O., & Chaffin, D. B. (1985). Ergonomic analysis of slip-resistance measurement devices. *Ergonomics, 28*(7), 1065–1080.

Barry, E. B., & Milburn, P. D. (1999). A footwear traction measuring device, International Society of Biomechanics. *Working Group on Functional Footwear Proceedings, Fourth Symposium on Footwear Biomechanics* (pp. 20–21). Canmore, Canada.

Braun, R., & Brungraber, R. J. (1978). A comparison of two slip-resistance testers. In C. Anderson & J. Seene (Eds.), *Walkway surfaces: Measurement of slip resistance* (pp. 49–59). ASTM STP 649, American Society for Testing and Materials.

Bring, C. (1964). *Friction and slipping*. Stockholm, Sweden: The National Swedish Council for Building Research.

Bring, C. (1982). *"Testing of slipperiness", Document D5*. Stockholm, Sweden: The National Swedish Council for Building Research.

Brungraber, R. J. (1976). An overview of floor slip-resistance research with annotated bibliography. National Bureau of Standard, NBS Technical Note 895, U.S. Department of Commerce, National Technical Information Service, PB-248 985, January, Washington.

Chaffin, D. B., Woldstad, J. C., & Trujillo, A. (1992). Floor/shoe slip resistance measurement. *Journal of American. Industrial Hygiene Association, 53*(5), 283–289.

Chang, W. (1998). The effects of surface roughness on dynamic friction between neolite and quarry tile. *Safety Science, 29*(2), 89–105.

Chang, W. (1999). The effects of surface roughness on the measurement f slip resistance. *International Journal of Industrial Ergonomics, 24,* 299–313.

Chang, W. R. (2001). The effects of surface roughness and contaminant on the dynamic friction of porcelain tile. *Applied Ergonomics, 32*(2), 173–184.

Chang, W. R. (2002). The effects of surface roughness and contaminants on the dynamic friction between porcelain tile and vulcanized rubber. *Safety Science, 40*(7–8), 577–591.

Chang, W. R., Kim, I. J., Manning, D. P., & Bunterngchit, Y. (2003). The role of surface roughness in the measurement of slipperiness. *Measuring slipperiness—Human locomotion and surface factors* (pp. 101–117). London: Taylor & Francis.

Cohen, H. H., & Compton, D. M. (1982). Fall accident patterns: Characterization of most frequent work surface-related injuries. *Professional Safety, 27*(6), 16–22.

Derler, S., Kausch, F., & Huber, R. (2008). Analysis of factors influencing the friction coefficients of shoe sole materials. *Safety Science, 46*(5), 822–832.

Demura, S., Kasuga, K., Sato, S., Sato, S., & Shin, S. (2013). Determination of persons at a high risk of falling in a population of healthy community-dwelling elderly Japanese. *International Journal of Gerontology, 7*(1), 13–16.

Dias, R. C., Freire, T. F., Santos, E. G. S., Vieira, R. A., Dias, J. M. D., & Perracini, M. R. (2011). Characteristic associated with activity restriction induced by fear of falling in community-dwelling elderly. *Revista Brasileira de Fisioterapia, 15*(5), 406–413.

DiDomenico, A., McGorry, R. W., & Chang, C. C. (2007). Association of subjective ratings of slipperiness to heel displacement following contact with the floor. *Applied Ergonomics, 38*(5), 533–539.

English, W. (1990). Improved tribometry on walking surfaces. In B. E. Gray (Ed.), *Slips, stumbles, and falls: Pedestrian footwear and surfaces, ASTM STP 1103* (pp. 73–81). Philadelphia: American Society for Testing and Materials.

Grönqvist, R. (1995). Mechanisms of friction and assessment of slip resistance of new and used footwear soles on contaminated floors. *Ergonomics, 38*(2), 224–241.

Harris, G. W., & Shaw, S.R. (1988). Slip resistance of floors: Users' opinions, Tortus instrument readings and roughness measurement. *Journal of Occupational Accidents*, *9*(4), 287–298.

Hausdorff, J. M., Rios, D. A., & Edelber, H. K. (2001). Gait variability and fall risk in community-living older adults: A 1-year prospective study. *Archives of Physical Medicine and Rehabilitation*, *82*(8), 1050–1056.

Health and Safety Executive (HSE). (2013). *Slips and trips*. UK.

Hornbrook, M. C., Stevens, V. J., Wingfield, D. J., Hollis, J. F., Greenlick, M. R., & Ory, M. G. (1994). Preventing falls among community-dwelling older persons: Results from a randomized trial. *Gerontologist*, *34*(1), 16–23.

Julian Robinson, J. (2016). *Moment Pope Francis, 79, FALLS OVER during Mass in front of a TV audience of millions while visiting Poland's holiest site*. Available at http://www.dailymail.co.uk/news/article-3712379/Moment-Pope-Francis-FALLS-Mass-TV-audience-half-million-people-visiting-Poland-s-holiest-site.html

Kim, I. J. (1996a). Tribological concepts for the investigation of the pedestrian slipping and falling accidents—Part I. *International Occupational Injury Symposium*, Sydney, Australia.

Kim, I. J. (1996b). Tribological approach for the analysis of pedestrian slip hazard—II. *Proceedings of the '96 Spring Conference of K.I.I.E.* (pp. 279–285), Soul, Korea.

Kim, I. J. (1996c). Microscopic investigation to analyze the slip resistance of shoes. *Proceedings of the 4th Pan Pacific Conference on Occupational Ergonomics* (pp. 68–73). Taiwan, ROC.

Kim, I. J. (2002). A pilot study on the measurements of heel contact areas for wear assessment. *XVI International Annual Occupational Ergonomics and Safety Conference*. Toronto, Canada, CD-Rom.

Kim, I. J. (2003a). Observation of the contact areas of the heel surface during dynamic slip resistance measures. *15th Triennial Congress of the International Ergonomics Association, IEA 2003, 7th Ergonomic Society of Korea/Japan Ergonomic Society Joint Conference*, Seoul, Korea, CD-Rom.

Kim, I. J. (2003b). A novel study on the correlation of the characteristics of contact area and average slope angle with dynamic friction coefficients. *15th Triennial Congress of the International Ergonomics Association, IEA2003, 7th Ergonomic Society of Korea/Japan Ergonomic Society Joint Conference*. Seoul, Korea, CD-Rom.

Kim, I. J. (2004a). Development of a new analyzing model for quantifying pedestrian slip resistance characteristics: Part I—Basic concepts and theories. *International Journal of Industrial Ergonomics*, *33*(5), 395–401.

Kim, I. J. (2004b). Development of a new analyzing model for quantifying pedestrian slip resistance characteristics: Part II—Experiments and validations. *International Journal of Industrial Ergonomics*, *33*(5), 403–414.

Kim, I. J. (2005). A new understanding on the shoe wear mechanism and its significance on slip resistance property. *Contemporary Ergonomics* (pp. 503–508). Chippenham, Wiltshire, Great Britain: Taylor & Francis, Antony Rowe Ltd.

Kim, I. J. (2006a). The current hiatus in fall safety measures. In W. Karwowski (Ed.), *International encyclopedia of ergonomics and human factors-2005* (pp. 2572–2576). LLC, USA: Taylor & Francis Group.

Kim, I. J. (2006b). A new paradigm for characterizing slip resistance properties. In W. Karwowski (Ed.), *International encyclopedia of ergonomics and human factors-2005* (pp. 2735–2740). LLC, USA: Taylor & Francis Group.

Kim, I. J. (2015). Wear observation of shoe surfaces: Application for slip and fall safety assessments. *Tribology Transactions*, *58*(3), 407–417.

Kim, I. J. (2016). Identifying shoe wear mechanisms and associated tribological characteristics: The importance for slip resistance evaluation. *Wear, 360–361,* 77–86.

Kim, I. J., Hsiao, H., & Simeonov, P. (2013). Functional levels of floor surface roughness for the prevention of slips and falls: Clean-and-dry and soapsuds-covered wet surfaces. *Applied Ergonomics, 44*(1), 58–64.
Kim, I. J., & Nagata, H. (2008). Research on slip resistance measurements—A new challenge. *Industrial Health, 46*(1), 68–78.
Kim, I. J., & Smith, R. (1998). A study of the comparative geometry mating between the surfaces of the shoe and floor in pedestrian slip resistance measurements. *The 5th Pan-Pacific Conference on Occupational Ergonomics* (pp. 34–37). Kitakyushu, Japan.
Kim, I. J., & Smith, R. (1999). The relationship between wear, surface topography characteristics and coefficient of friction as a means of assessing the slip hazards. *The 2nd Asia-Pacific Conference on Industrial Engineering and Management Systems (APIEMS '99)* (pp. 155–161). October, Ashikaga, Japan.
Kim, I. J., & Smith, R. (2000). Observation of the floor surface topography changes in pedestrian slip resistance measurements. *International Journal of Industrial Ergonomics, 26*(6), 581–601.
Kim, I. J., & Smith, R. (2003). A critical analysis of the relationship between shoe-floor wear and pedestrian/walkway slip resistance. In M. I. Marpet & M. A. Sapienza (Eds.), *Metrology of pedestrian locomotion and slip resistance* (pp. 33–48). West Conshodocken, Pennsylvania, USA: ASTM International: STP 1424.
Kim, I. J., Smith, R., & Nagata, H. (2001). Microscopic observations of the progressive wear on the shoe surfaces which affect the slip resistance characteristics. *International Journal of Industrial Ergonomics, 28*(1), 17–29.
Leclercq, S., & Saulnier, H. (2002). Floor slip resistance changes in food sector workshops: Prevailing role played by fouling. *Safety Science, 40*(7–8), 659–673.
Li, K. W., Chang, W. R., Leamon, T. B., & Chen, C. J. (2004). Floor slipperiness measurement: Friction coefficient, roughness of floors, and subjective perception under spillage conditions. *Safety Science, 42*(6), 547–565.
Liberty Mutual. (2003). 2003 Liberty Mutual Workplace Safety Index: Identifies the direct costs and leading causes of workplace injuries. *Liberty Mutual Research Institute for Safety, Fall.* Available at file:///H:/2-IJKIM/Papers/2015-4-Book/2003%20Liberty%20Mutual%20Workplace%20Safety%20Index.pdf
Manning, D. P., Jones, C., Rowland, F. J., & Roff, M. (1998). The surface roughness of a rubber soling material determines the coefficient of friction on water-lubricated surfaces. *Journal of Safety Research, 29*(4), 275–283.
Martin, G. (1992). Practical slip-resistance testing. *Journal of Occupational Health Science - Australia NZ, 8*(6), 505–510.
Maurer, R. (2012). Fatal work injuries decline slightly in 2011. *Society for Human Resource Management (SHRM). Sept.* Available at http://www.shrm.org/hrdisciplines/safetysecurity/articles/pages/fatal-work-injuries-decline-2011.aspx
Mills, R., Dwyer-Joyce, R. S., & Loo-Morrey, M. (2009). The mechanisms of pedestrian slip on flooring contaminated with solid particles. *Tribology International, 42*(3), 403–412.
National Research Council. (1961). *Causes and measurement of walkway slipperiness*. Federal Construction Council, Technical Report No. 43, Washington, DC: National Academy of Sciences—NRC, Publication 899.
Ozanne-Smith, J., Guy, J., Kelly, M., & Clapperton, A. (2008). *The relationship between slips, trips and falls and the design and construction of buildings*. Monash University Accident Research Centre. Report No. 281.
Peel, N., Bell, R. A. R., & Smith, K. (2008). *Queensland stay on your feet® community good practice guidelines—Preventing falls, harm from falls and promoting healthy active ageing in older Queenslanders*. Queensland Health, Brisbane. Available at https://www.health.qld.gov.au/stayonyourfeet/documents/33383_full.pdf
Perez-Jara, J., Olmos, P., Abad, M. A., Heslop, P., Walker, D., & Reyes-Ortiz, C. A. (2012). Differences in fear of falling in the elderly with or without dizziness. *Maturitas, 73*(3), 261–264.

Proctor, T. D., & Coleman, V. (1988). Slipping and tripping accidents and falling accidents in Great Britain—Present and future. *Journal of Occupational Accidents, 9*(4), 269–285.

Redfern, M. S., Cham, R., Gielo-Perczak, K., Grönqvist, R., Hirvonen, M., Lanshammar, H., et al. (2001). Biomechanics of slips. *Ergonomics, 44*(13), 1138–1166.

Rowland, F. J., Jones, C., & Manning, D. P. (1996). Surface roughness of footwear soling materials: Relevance to slip-resistance. *Journal of Testing and Evaluation, 24*(6), 368–376.

Rubenstein, L. Z., & Josephson, K. R. (2006). Falls and their prevention in elderly people: What does the evidence show?. *The Medical Clinics of North America, 90*(5), 807–824.

Safe Work Australia. (2012). *Slips and trips at the workplace fact sheet.* Available at http://www.safeworkaustralia.gov.au/sites/swa/about/publications/Documents/659/Slips%20and%20Trips%20Fact%20Sheet.pdf

Safe Work Australia. (2013). *"Key work health and safety statistics", Australia.* Available at http://www.safeworkaustralia.gov.au/sites/SWA/about/Publications/Documents/758/Key-WHS-Statistics-2013.pdf

Washington State Department of Labour and Industries (WSDLI). (2010). Slips, trips and falls. Washington, USA.

Whitney, S. L., Marchetti, G. F., Ellis, J. L., Otis, L. (2013). Improvements in balance in older adults engaged in a specialized home care falls prevention program. *Journal of Geriatric Physical Therapy, 36*(1), 3–12.

World Health Organization (WHO). (2012). *"Fact sheet 344: Falls", World Health Organization.* Available at http://www.who.int/mediacentre/factsheets/fs344/en/

Chapter 2
Pedestrian Fall Incidence and Slip Resistance Measurements

2.1 Brief Overview of Slip and Fall Incidences

As discussed in Chap. 1, slip and fall incidences are widely studied owing to their prevalence and associated costs in terms of human suffering and economic burden around the world. The incidents are categorized into falls from an elevation and same level falls. Falls from an elevation are those where the point of contact is below the level of the original supporting surface of the faller. Falls from the elevation are one of the leading causes of workplace fatalities and traumatic brain injuries on the job. These incidents create a substantial risk that an employee or worker will become momentarily or permanently disabled.

Whilst the same level falls are those where the point of contact is on the same level or above the original supporting surface of the faller. Falls from the elevation are considered more likely to lead to severe injuries but the same level falls occur far more frequently. In some cases, factors intrinsic to the pedestrians are solely responsible for the same level falls. But more often, these falls are at least partially induced by environmental factors. For example, trips occur during walking when the leading foot is arrested by an obstruction which interrupts the smooth movement of the body's centre of mass.

Slips are a more common cause of the same level falls, contributing to up to 85% of all fall-related occupational injuries (Courtney et al. 2013). They occur when the underfoot conditions induce a sudden loss of grip because the coefficient of friction (COF) between the shoe heel (or barefoot) and the floor surface is insufficient to resist the forces at the point of contact (Leamon 1992). Foot slips and trips are injurious when they result in harmful loading of body tissues as a result of a sudden release in energy (Grönqvist et al. 2001).

A number of gait studies from the literature have confirmed that slips typically occur either when the trailing foot is pushing off (toe-off) or when the leading foot contacts the ground (heel strike) (Perkins and Wilson 1983; Redfern et al. 2001).

I.-J. Kim, *Pedestrian Fall Safety Assessments*,
DOI 10.1007/978-3-319-56242-1_2

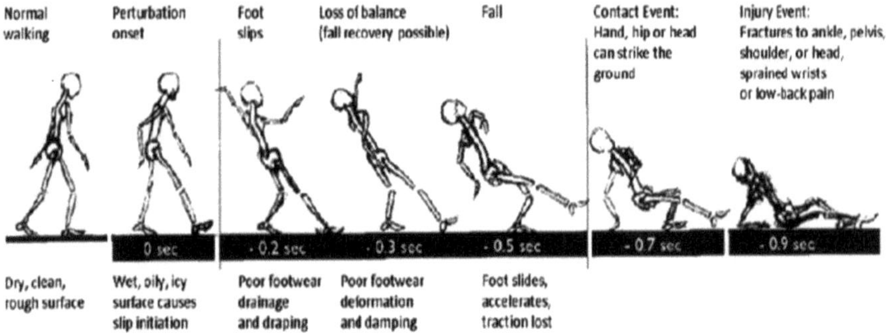

Fig. 2.1 Typical event sequences for slips and falls (Grönqvist et al. 2001)

At the phase of toe-off, the forces generated at the foot are used to propel the body's centre of mass forward after the majority of weight has already been transferred to the contralateral foot.

At the stage of heel strike, the vertical component of weight is transferred onto the leading foot. As weight is being transferred the body is inherently unstable, relying on safe planting of the striking foot for momentary stabilisation. Accelerations at the heel during heel strike can, therefore, lead to balance loss and potential falls. As such, slips at toe-off are less hazardous than those at heel strike (Leamon 1992).

Slip resistance is a term used to describe properties of underfoot surfaces and footwear that resist the tendency to slide relative to one another (Grönqvist et al. 2001; ASTM F1637 2013). The provision of adequate slip resistance is important in reducing the risk of slip and fall incidents. Figure 2.1 illustrates the typical sequence of normal walking followed by slip initiation at heel strike due to inadequate slip resistance. This then leads to balance loss, fall, and injury, accordingly.

2.2 Injuries Owing to Slips and Falls

Fall incidents from slips or trips are hazardous because they cause to injuries (Baker and Harvey 1985; Blake et al. 1988). Common injuries include sprains and fractures often affecting the wrists, pelvis and lower extremities. Hip fractures involving elderly people are of great concerns in falls. Whilst the same level fall incidents are hardly life-threatening, patients are at increased risk of premature death for several years following a hip fracture (Abrahamsen et al. 2009). Injuries resulting from falls have also been found to greatly affect the activity level and lifestyle of older people.

In the workplace, the injury frequencies from slips and falls also affect the older workers more than younger ones and can impact differently between men and women employees. Nenonen (2013) used data mining techniques to analyse occupational slips, trips and falls from the Finnish occupational accidents and diseases statistics database. This study found out that slip, trip, and fall incidents incurred 66% in the male workers and 34% in the female ones, respectively.

Layne and Landen (1997) studied hospital emergency department records of 136,985 work-related injuries involving workers over the age of 55 across the United States in 1993. Fall-related incidents accounted for 26.3% of all non-fatal injuries, of which same-level falls were the most common (55.9%). Women experienced more same-level falls (67.9%) than men. Falls were also the leading cause of hospitalization (43.9%) and fractures and dislocations sustained as a result of falls accounted for the greatest proportion of hospitalizations (36.1%).

2.3 Improvements of Fall Prevention

High traction or slip resistant floors and floor coverings and proper maintenance procedures have been found to reduce slips and falls. Specialized flooring materials for increased grip functioning including abrasive paints and tactile strips have been shown to be useful (Bell 1997). Cautious and immediate elimination of spills and other contaminants on floor surfaces as well as removal of any obstructions are also important for decreasing falls (Weisberger 1994). Other environmental changes that can reduce fall risks include the application of warning signs (Gadomski 1998) and improving lighting conditions (Maynard 2006).

Bentley (2009) described the importance of perception and cognition of hazards in the prevention of fall incidents. His study of construction workers in New Zealand found that in 75% of cases, workers did not notice the fall hazard prior to the incident. Typical reasons for having not perceived hazards a priori include distraction or divided attention or the hazard was obstructed from view.

In a review of the literature, Bell (1997) suggested that aside from improving slip resistance of floor surfaces, the other effective way to prevent slip and fall incidents in the workplace was through footwear. Targeting improvements in footwear for increasing slip resistance has the potential to impact falls on a large scale as evidenced by various cases in indoor industrial environments. Verma et al. (2011) showed that adopt of slip resistant shoes from 36 limited-service restaurants across the US reduced slipping rates by 54%. A further analysis of the slip resistant footwear in these restaurants showed that older workers and women were more likely to use the slip resistant footwear (Verma et al. 2011).

Courtney et al. (2013) studied the relationship between the subjective perception of slipperiness and the risk of slipping in the workplace. They found out that when the perception of slipperiness (rated on a 4-point scale) was assessed across restaurants, the awareness of slipperiness was strongly associated with rates of slipping. For every 1-point increase in the mean rating of slipperiness, there was a 2.71 times increase in the rate of slipping.

Staal et al. (2004) studied the effectiveness of positive-grip (or high traction) shoe covers in a hospital setting where the rates of slipping incidents had increased as a result of changing the hospital floors from carpet to porcelain tile. In the initial phases of this study, they found out that majority of the fall incidents had occurred whilst employees were helping patients to transfer from shower chairs on wet floors. From a small sample of shoes, they concluded that high traction shoe covers were effective in preventing fall incidents.

Many other studies from the literature have also indicated a need for specialised footwear and many researchers have suggested the use of spikes or studs for added traction on risky environments. However, there remains very little evidence to show how such anti-slip devices actually impact slip and fall risks. Pedestrians continue to experience high rates of slips and falls despite using them (Rolfsman et al. 2012). Similarly, whilst certain outsole materials and tread designs have been touted to potentially reduce falls, highly effective designs for shoe sole/heel treads have not been established. In order to find effective footwear solutions for preventing slips and falls, an understanding of the interacting factors between the human, the footwear, and the environment is necessary (Gao and Abeysekera 2004).

2.4 Factors Influencing Pedestrian Fall Incidence

Pedestrian slip and fall incidents are caused by a number of reasons. As briefly illustrated in Fig. 2.2, multi-factorial risk factors seem to, directly and indirectly, affect the incidence. These risk factors signify that fall safety assessment is a highly complex area to study, where the likelihood of a slip and fall is a function of a variety of elements such as surfaces (types, materials, and surface finishes), environmental conditions (dry or lubricated), and individual users (physical conditions and footwear).

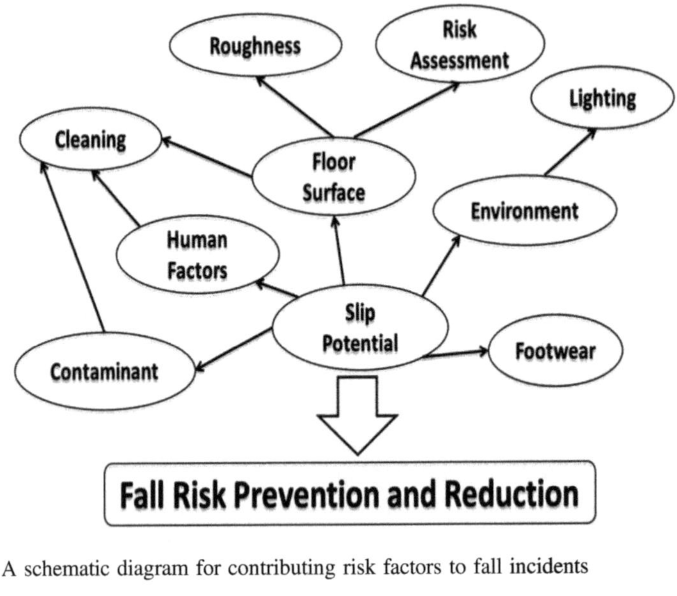

Fig. 2.2 A schematic diagram for contributing risk factors to fall incidents

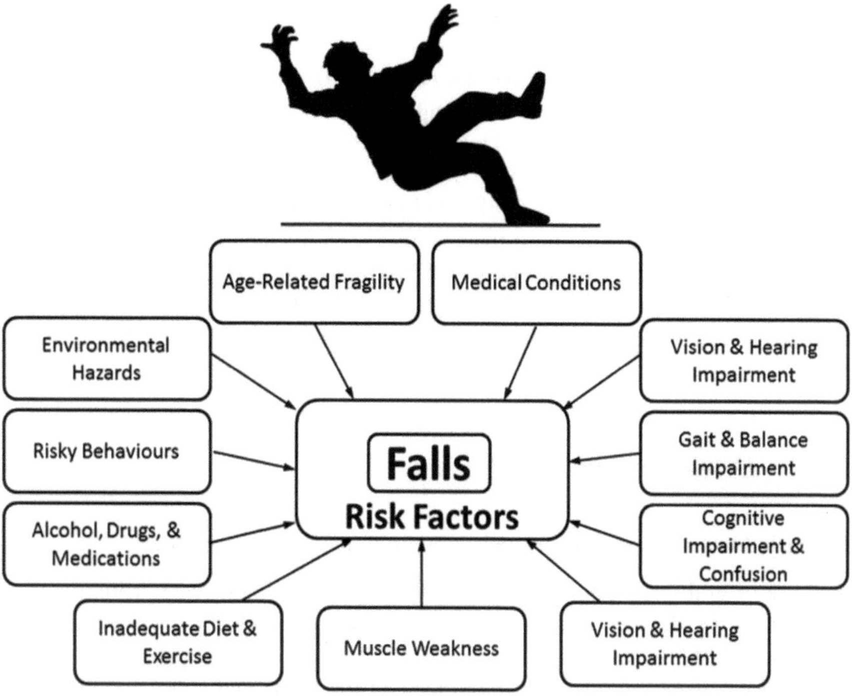

Fig. 2.3 A schematic diagram for Intrinsic and extrinsic fall risk factors

As shown in Fig. 2.3, slip and fall incidents are a continuing problem and frequently caused by a combination of risk factors that are specific to physiological (intrinsic factors) and environmental (extrinsic factors) conditions (Pearson and Coburn 2011; Titler et al. 2011).

If these risk features are identified, then it would be beneficial to develop appropriate prevention strategies for the fall incidence.

The following is a brief summary of the recent studies on intrinsic and extrinsic fall risk factors:

2.4.1 Intrinsic Fall Risk Factors

Intrinsic risk factors are mainly related to physiological issues such as

- Age-related changes (weakened vision, mobility, and gait issues) (von Schroeder et al., 1995; Evans et al. 2001; Hathaway et al. 2001; Krauss et al. 2005, 2007; Currie 2008; Carroll et al. 2010; Tinetti and Kumar 2010).
- Arthritis (Rubenstein and Josephson 2006; Peeters et al. 2009).
- Chronic illness (The Joint Commission 2007).

- Confusion and dizziness (Evans et al. 2001; Tzeng 2010b; Gray-Miceli et al. 2012).
- Depression (Huang et al. 2005; Yu et al. 2009).
- Length of hospital stay, fear of falling, and history of falls (Hitcho et al. 2004; The Joint Commission 2007; Dykes et al. 2009; Tzeng 2010a, b; Pearson and Coburn 2011; Titler et al. 2011; Boushon et al. 2012).
- Lower extremity weakness (Helbostad et al. 2010; Hatton et al. 2013).
- Polypharmacy (five or more medications) (Hartikainen et al. 2007; The Joint Commission 2007; Rhalimi et al. 2009; Titler et al. 2011).
- Status of activity of daily living (Zijlstra et al. 2007; Leung et al. 2010).
- Urinary incontinence (Hitcho et al. 2004; Dykes et al. 2009; Tzeng 2010a).

2.4.2 *Extrinsic Fall Risk Factors*

Extrinsic factors are mainly related to environmental issues such as

- Lack of grab bars, poor condition of floor surfaces, inadequate or improper use of assistive devices (Donald et al. 2000; Agostini et al. 2001; Pearson and Coburn 2011).
- Surface finishes and wear developments of the floors and shoes and their interactive effects on slip resistance performance (Kim 1996a, b, c, d, 2004a, b, 2006a, b, 2015a, b, c, d, 2016a, b; Kim and Smith 1998a, b, 1999, 2000, 2001a, b, c; Kim et al. 2001, 2003, 2013; Chang et al. 2003a, b; Kim and Nagata 2008a, b).
- These can be minimised through improved design and installation practices, better cleaning maintenance practices, safety audits, remedial policies, and mandatory legislation (Kim et al. 2013; Kim 2015b). Shoes may be considered as an extrinsic factor since inappropriate or excessively worn heels/soles may be the prime cause of an accident (Kim 2015a, c, 2016b).

Recognizing that widening the perspectives may better define the role of multi-factors in detection and prevention of fall incidence, but this book is mainly focused on exploring environmental (floorings and shoes) correlations for slip and fall incidents under a range of environmental conditions as a starting point. However, following sections from 2.2 to 2.5 briefly overview human walking, gait and its impacts on fall incidents, gait analysis and fall risk prediction, respectively.

2.4.3 *Mechanics of Human Walking*

Human walking (or gait) is a method of locomotion in which the body weight (or centre of gravity) is carried alternately by the right and left foot (Skoyles 2006). Interestingly, every individual has a unique gait pattern. Human's gait can be largely affected by injury or disease processes. By evaluating the gait patterns of an individual, a clinician, rehabilitation scientist, and/or physiotherapist can determine specific weaknesses and develop recovery programs to address detailed issues and problems.

Objective assessments of human gait parameters such as stride length, heel velocity, and slip distance during ambulation under normal and abnormal conditions have been conducted by many investigators (Herman et al. 1976; Crowinshield et al. 1978). These gait parameters also have been used to develop slip resistance measuring devices and apparatus for tribo-physical studies amongst the shoe heels, floor surfaces, and slip resistance properties. By studying human gait and gait-related parameters, more practical measurements and analyses for the pedestrian fall safety can be obtained (Perkins 1978; Strandberg and Lanshammar 1981).

2.5 Human Gait and Its Impacts on Fall Incidents

Human gait depends on several factors such as body weight, body shape, and age and is unique to an individual. The motion in which the foot contacts the floor is equally exclusive and can be both positive and negative with respect to the overall direction of motion (Whittle 1999).

The gait cycle is the term describing an ambulatory phase of walking or running. It is rather complex with each gait pattern being unique to each individual. There are many reasons that our gait patterns are different amongst us, but there are still core components that can be measured and evaluated by biomechanists, podiatrists, clinicians, or researchers.

There are four main phases of gait stages during foot-floor contact: namely, impact (initial contact), foot-flat, propulsion (midstance and heel lift), and toe-off. Figure 2.4 shows a gait cycle during normal walking.

As illustrated in Fig. 2.4, one leg is in the stance phase whilst the other is in the swing phase. Muscles must contract to counterbalance the forces of gravity, to offer acceleration or deceleration to momentum forces, and to overcome the resistance of walking surfaces.

Slips typically occur at the initial impact stage, known as a heel strike, where the foot generates a converging wedge and the contact area is small. In the case of fluid contaminants, this promotes the formation of fluid films and hydrodynamic lubrications. However, the small contact areas between the foot heel (or shoe heel) and the floor surface and relatively large sizes of contaminant particles (or pollutants) may significantly affect traction (or friction) properties and reduce slip resistance performance.

Redfern et al. (2001) suggest that the peak load (expressed as impact force per unit body weight) is approximately 10 N/kg (or roughly equal to body weight) occurring at 25% into the stance phase during walking on a horizontal surface. However, the foot reaches the 'foot-flat' (FF) position approximately 15% into the stance phase at which point the reaction force is around 8.5 N/kg (see, Fig. 2.5).

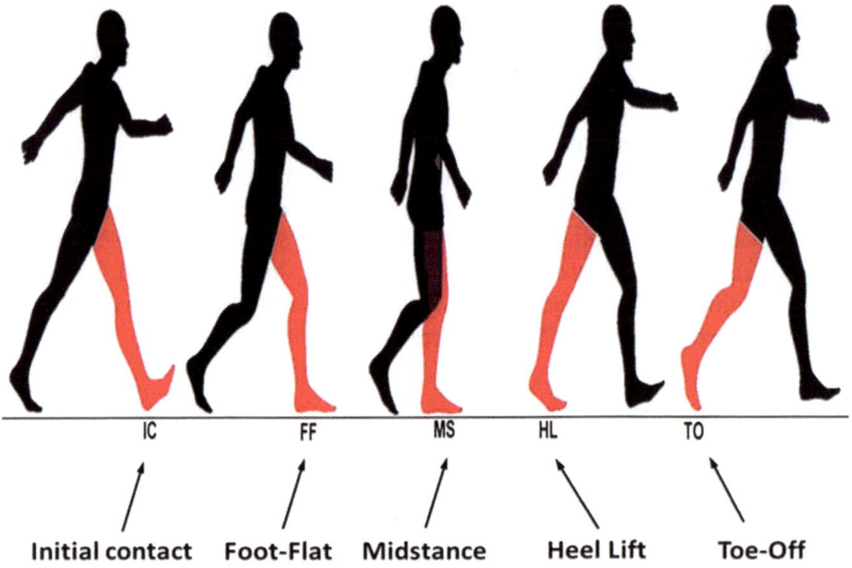

Fig. 2.4 Phases of gait cycle: impact (initial contact), foot-flat, propulsion (midstance and heel lift), and toe-off

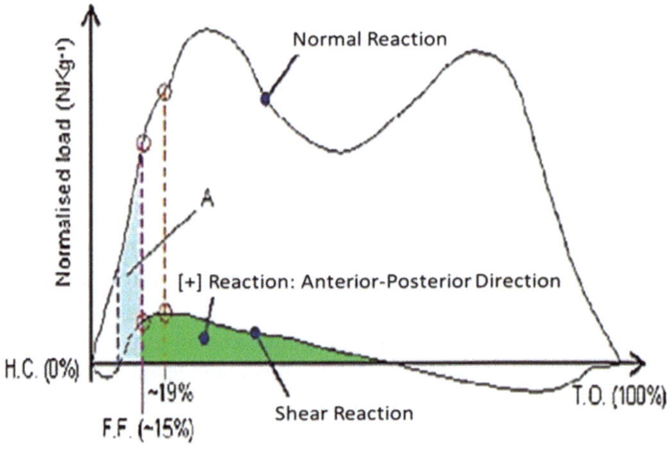

Fig. 2.5 Relative variation of normal and shear reactions over a single step (Whittle 1999)

The position of the highest fall risk is likely to be just prior to this point when the foot is still inclined to the floor. The pressure generated seems to depend upon the shoe contact area with the ground surface (highly shoe and gait dependent). Despite the importance of this contact mechanism during the gait phase, any detailed information on this issue is hardly found in the literature.

The maximum shear force of approximately 1.5 N/kg occurs at a position (19%) into the stance (Redfern et al. 2001), but the critical position in the stance where the shear to normal load ratio is the largest and contact area the smallest. is likely to occur within the shaded region (*A*) of Fig. 2.5 (before FF position), when the shear stress ranges from 0 to around 1.25 N/kg.

2.6 Observation of Human Gait

Walking is one of the most fundamental actions in our daily life. Regardless of its severity, gait injury can affect the overall quality of day-to-day living. Clinicians and researchers are interested in examining gait behaviours to characterise altered movement patterns due to gait impairment. Gait imperfection can be caused by musculoskeletal disorders and/or neurological diseases (Ashton-Miller 2005), but even healthy individuals can have difficulty in walking due to injury from an unforeseen accident. Many injured individuals can fully recover from trauma, surgery, or sudden pathology, but significant numbers of injured people also turn out with permanent damages in their walking after rehabilitation (Wade et al. 1987; Burdett et al. 1988; von Schroeder et al. 1995).

The main objective of human gait analysis is to identify and uncover underlying gait pathologies or impairment and develop rehabilitation programs (Park 2012). Hence, better understandings on gait patterns and movements may help clinicians and researchers to develop a strategy to prevent injury or long-term disability. In addition, observation of movement patterns from body segments and joints during walking can reveal important information on gait performance. Deviations from normative movement patterns can be an indicator of nerve damages, injuries, anatomical abnormalities, and other neurological or musculoskeletal problems (Park 2012).

Gait analysis is based on an assumption that gait abnormality is due to underlying neurological or musculoskeletal disorders. In addition, the environmental hazard can affect gait behaviour, resulting in abnormal gait patterns. Thus, quantitatively assessing deviations from normative gait patterns can indirectly provide insights into the mechanisms behind neurological and/or musculoskeletal illnesses and adaptations of a challenging environment that affects joint movements during walking or running.

Gait assessments mainly consist of two areas: qualitative observation and quantitative approach. Qualitative measures include visual inspection of gait behaviours (Perry and Burnfield 2010). The Hoehn and Yahr scale described the severity of symptoms of Parkinson's disease (PD) (Hoehn and Yahr 1998), and the Berg Balance scale measured static and dynamic balance ability (Berg et al. 1992). However, the usage of qualitative measures is limited because this assessment approach cannot provide detailed information on gait behaviours.

Moreover, the qualitative measurements are based on observers' subjective opinions so they are often not consistent and unable to capture the .small deviations of specific gait patterns in comparison to healthy individuals' ones. Therefore, information and judgment based on a clinician's subjective observation may be insufficient to accurately diagnose a patient's gait condition.

To address the deficiencies of the qualitative assessment, quantitative gait analysis can be adopted with laboratory equipment and instruments such as computerised motion capture systems and force platforms. These quantitative measurements enable clinician and researchers to conduct very detailed gait analyses. Continuous movement patterns can reveal important and detailed information on the gait analysis for a subject through deviations from normal gait patterns.

In the quantitative analysis, conventional techniques to characterize human gait behaviours have mainly focused on examining discrete events during a specific task-related motion, such as heel strike and toe-off during walking, or on using univariate statistical measurements such as a stride and step length, and the durations of stance and swing phases of gait (Danion et al. 2003; Diop et al. 2004; Hausdorff 2004; Knoll et al. 2004; Owings and Grabiner 2004; Schwartz et al. 2008).

The quantitative gait analysis has the ability to capture complex motion patterns produced throughout the gait cycles. Human gait is a spatiotemporally complex movement that involves interactions between multiple body segments and couplings across multiple joints. It is, therefore, crucial to assess complex motion patterns and correlated movements across multiple joints in order to detect physiological and neurological constraints, limitations, and injuries.

2.7 Gait Analysis and Fall Risk Prediction

Gait analysis offers reasonable tools to detect various population groups prone to falling. Gait disturbances and instabilities can result in increased slip and fall risks for older adults, people with PD, and workers in a range of industries.

Previous studies proposed that asymmetry, variability, complexity, and joint couplings during walking can be used as markers of slip and fall hazards in a variety of populations (Tiberio 1987; Blin et al. 1990; Hausdorff et al. 1998; Hamill et al. 1999; Frenkel-Toledo et al. 2005). Despite the detailed studies on gait and balance controls in fallers and non-fallers, specific gait behaviours that are critically associated with falls and fall predictions still remain unclear. New tools for investigating joint variability, symmetry, and coupling in gait motion patterns may provide added insight to this association.

Slip and slip-related fall incidents are responsible for many musculoskeletal injuries (Strandberg and Lanshammar 1981; Manning et al. 1988) and occur when the level of friction available at the shoe/floor interface is less than expected (Tisserand 1985; Leamon and Son 1989). If a low friction level is expected, gait adjustments can reduce the likelihood of a slip (Cham and Redfern 2002) and increased readiness can improve the likelihood of balance recovery if a slip occurs (Marigold and Patla 2002). The obvious benefits of these gait adjustments and increased readiness are fewer slip-related injuries. However, many real-world slip and fall incidents occur unexpectedly (Leclercq 1999), whereas most laboratory-based slip and fall simulations happen with some level of prior knowledge (Cham and Redfern 2002; Marigold and Patla 2002; Andres et al. 1992; You et al. 2001; Marigold et al. 2003; Brady et al. 2000; Bunterngchit et al. 2000; Hanson et al. 1999; Lockhart et al. 2003; Oates et al. 2004; Pai et al. 2003; Pavol et al. 2004; Hirvonen et al. 1994).

Although the recent studies have attempted to quantify differences between laboratory-based gait trials and real world gait events (Cham and Redfern 2002; Marigold and Patla 2002; Andres et al. 1992), their results showed that subjects who knew the floor was potentially slippery approached heel strikes with a flatter foot and a more vertical shank, even if asked to walk normally (Cham and Redfern 2002; Marigold and Patla 2002; Andres et al. 1992). Subjects also reduced their vertical and horizontal ground reaction forces and impulses during heel contacts, and thus reduced the friction necessary to walk without slipping (Cham and Redfern 2002; Marigold and Patla 2002; Andres et al. 1992).

Such gait adaptations have been cited as limitations in many slip and fall experiments (Perkins 1978; Kulakowski et al. 1989; Leamon and Son 1989; Andres et al. 1992; Hirvonen et al. 1994; Fendley and Marpet 1996; Marpet 1996; Brady et al. 2000; Bunterngchit et al. 2000) and may affect the external validity of some slip and fall studies (Leclercq 1999).

Prior knowledge encompasses both awareness of potential future slips and experiences gained from the past slips. Multiple slips are used in some studies and the recent experience gained from prior slips can be integrated with awareness to refine the gait adjustments. Thus, it is important to understand the effect of awareness on both experienced and inexperienced subjects to determine whether awareness or experience has a dominant effect.

Andres et al. (1992) varied both awareness and experience simultaneously and showed that the independent effects of these variables could not be discerned from their study. Cham and Redfern (2002) observed awareness-related changes in normal gait but pooled data from subjects with and without slip experience.

Marigold and Patla (2002) observed that the combined awareness and prior slip experience produced a “normal” gait different from that of both experienced and inexperienced subjects who knew for certain they would not slip.

Again, the independent effects of awareness and experience, and in particular the effect of awareness on inexperienced subjects, could not be discerned from their data. Therefore, detailed understandings on how prior knowledge affects the genesis of a slip is needed to better analyse slip and fall assessment results.

2.8 Measuring Devices for Slip Resistance Properties

A range of apparatuses and devices have been fabricated and developed to quantify frictional behaviours between the shoe and floor. The earlier studies related to slip and fall incidents were focused exclusively on the measurements of a COF quantity.

For friction measurements, about 70 different types of slip measuring testers and instruments are recorded in the literature and they differ substantially in their designs (Chang et al. 2001; Expert Panel 2004). In general, most slip resistance measuring devices and instruments fall into four categories: articulated strut, drag/towed-sled, pendulum, and other types, respectively. Following sub-section is the summary of those apparatuses for the measurements of slip resistance properties.

2.8.1 *Articulated Strut Devices*

This device consists of a weight attached to a shaft articulated at an angle. The angle of articulation increases until a shoe sole sample slips. The tangent of the angle at which the shoe sole slips is related to a static COF. The James Machine developed by Sidney James of Underwriters Laboratories in 1951 is a good example of the articulated strut devices. The James Machine is a laboratory apparatus for dry testing only (Fig. 2.6a). As an articulated strut class of tribometer, the James Machine applies a known constant vertical to a test pad (i.e., a leather pad when testing flooring materials), and then applies an increasing lateral force until a slip occurs.

The two NBS (National Bureau of Standards) testers: Brungraber Portable Slip Resistance Tester (Mark I and II) and Ergodyne Slip Resistance Tester are examples of the articulated strut device as well. Figure 2.6 shows photographic images for the James Machine and Brungraber Portable Slip Resistance Testers: Mark I and II.

2.8.2 *Drag and Towed-Sled Devices*

This type of device slides a weighted scale mounted with a shoe sole sample across a test surface. Such a device is pulled either manually or by a motor at an adjustable speed. It is the most common type of a slip meter due to its simplicity, portability, and ease of use. The COF is derived by dividing the force required to cause slippage by the weight of the sled. Some models measure only static friction and others do both static and dynamic frictions.

Figure 2.7 shows photographic images for examples of the drag and towed-sled type devices: the Horizontal Pull Slipmeter (HPS) and the Tortus Floor Friction Tester designed by British Ceramic Research Limited, respectively.

(**a**) James Machine

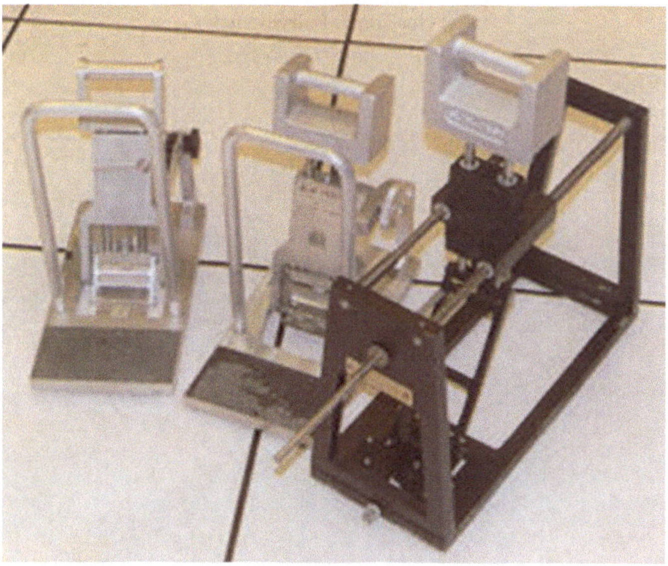

(**b**) Brungraber Mark I and II

Fig. 2.6 Photographic images for **a** the James Machine and **b** Brungraber Portable Slip Resistance Tester, Mark I and II

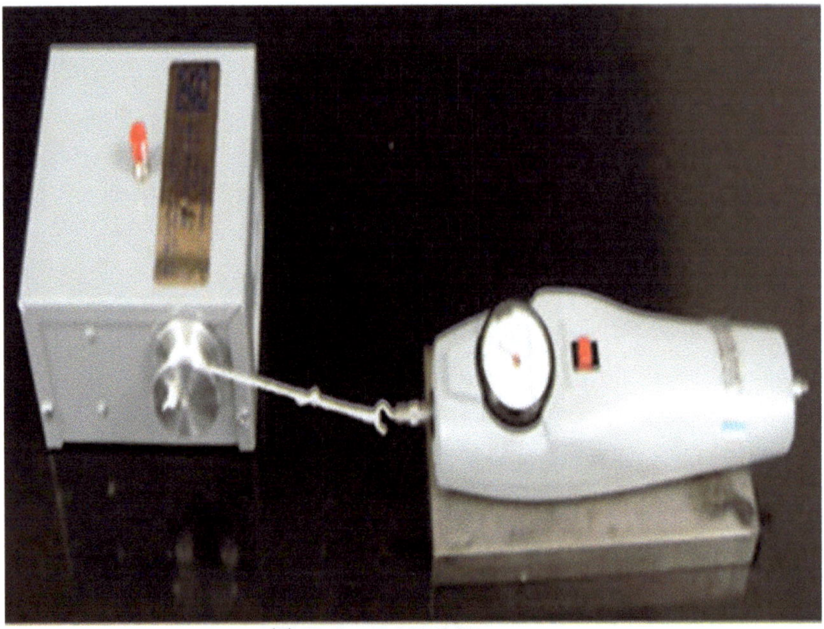

(**a**) Horizontal Pull Slipmeter

(**b**) Tortus Floor Friction Tester

Fig. 2.7 Photographic images for **a** the horizontal pull slipmeter (HPS) and **b** the Tortus floor friction tester designed by British Ceramic Research Limited

The Horizontal Pull Slipmeter (HPS; ASTM F 6098-79), the Bigfoot designed by Safety Sciences, Inc. (Lockhart 1997), Dynamometer Pull-Meter (ASTM C 1028-84), CEPBT Skidmeter (Majcherczyk 1978), and the Tortus Floor Friction Tester designed by British Ceramic Research Limited (Ceram Research) are the drag type devices which are most frequently referenced in the literature.

2.8.3 *Pendulum Type Devices*

This type of device is represented by the Sigler Device, the British Portable Skid Test (TRRL Pendulum Tester), Wessex Universal Tester, Small Pendulum Impact Tester and measures only dynamic friction. A shoe sole material is attached to a mechanical foot that is impacted onto and swept over a floor surface at a relatively high speed. Dynamic friction is measured by the energy loss of the pendulum at the beginning and the end of a swing.

One of the earliest comprehensive studies on COF measurements was conducted by Sigler et al. (1948). They developed a pendulum type of device to measure the COF between a slider of leather and a type of rubber and 23 different underfoot surfaces against dry and wet conditions.

Figure 2.8 shows photographic images for the Horizontal Pull Slipmeter, Sigler Pendulum Tester, and English XL Tribometer, respectively.

2.8.4 *Other Type Devices*

Braun and Roemer (1974) developed a device to determine static and dynamic friction properties between a shoe heel/sole material and floor surface. The objective of this development was to investigate the effect of polishes on COFs. Five different flooring materials and eight commercially available polishes were used with their device. The friction measurements were conducted before and after the treatments of polishing. It was interesting to note that the COFs increased in almost every case after the surfaces were treated with polishes. The average of static COFs on the untreated floor surfaces was 0.32, and the average COF was increased to 0.45 against the treated floor ones. For the case of dynamic COFs, the average dynamic COF was increased from 0.28 to 0.42. Therefore, the polishing process increased slip resistance performance by an average of over 30%.

Reed (1975) designed and constructed a portable type friction measuring device to test industrial working surfaces and named this device as "Universal Friction Testing Machine (UFTM)" (see, Fig. 2.9). This device was used to determine the values of COF between two shoe materials: leather and rubber and three-floor surfaces: vinyl (asbestos reinforced), oak (across the grain), and oak (with the grain).

Each floor material was tested against the leather and rubber specimens under dry conditions at different velocities from static to dynamic of 152 cm/s. A table of

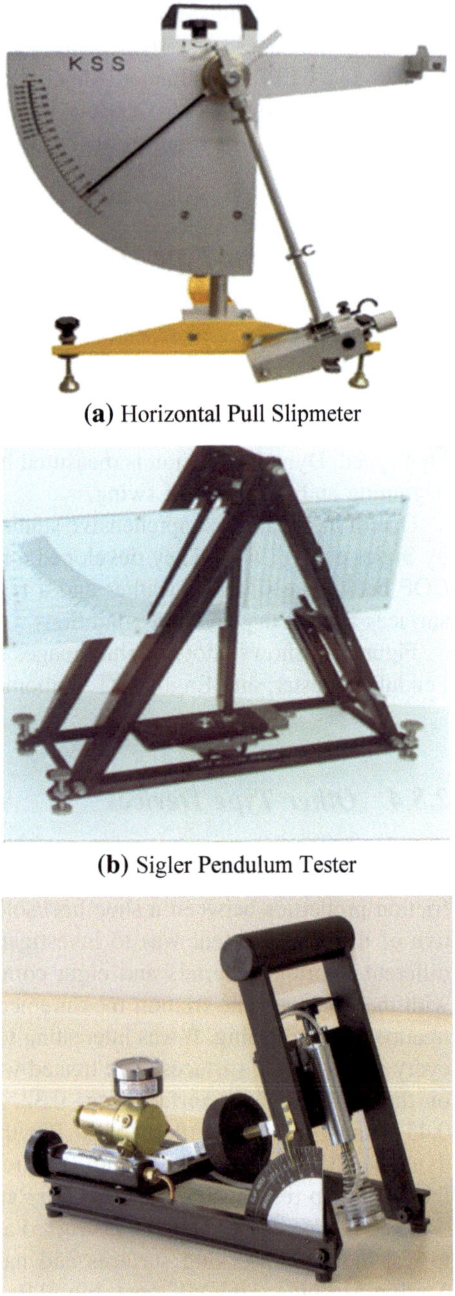

Fig. 2.8 Photographic images for **a** the horizontal pull slipmeter, **b** Sigler pendulum tester, and **c** English XL Tribometer, respectively

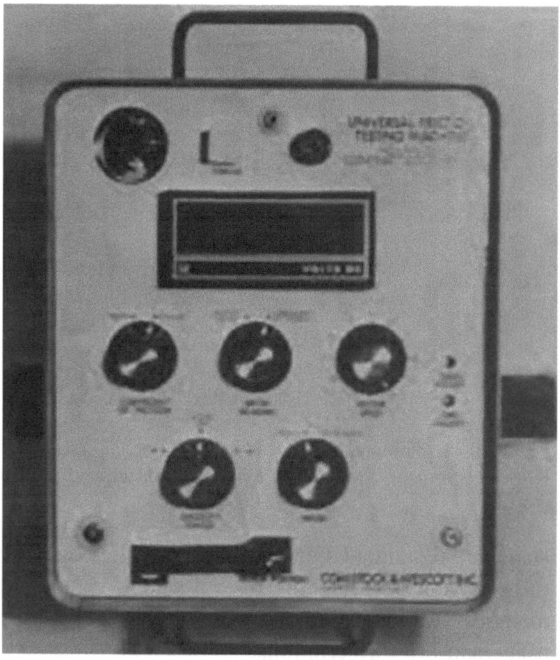

Fig. 2.9 A photographic image for the universal friction testing machine which was designed expressly to enable the National Institute for Occupational Safety and Health to measure COFs

COFs was produced at various slip speeds. However, it should be pointed out that the vertical forÎ applied to this device could not provide a good simulation of human walking and suffered from serious reliability problems. As a result, this device was never commercially produced.

Other types of slip resistance measuring devices also include the Multicomponent Quartz Force Plate (FP) manufactured by Kistler Instrument, Switzerland. The Quartz FP device measures the magnitude and direction of the ground to foot reaction force in three component forces (mediolateral, anterior-posterior, and vertical), static and dynamic COFs as well as the point of force application or centre of pressure during foot contact. The force is recorded when the subject walks over the floor sample mounted on the plate top. Detailed descriptions of the mechanisms of the FP can be found in the literature (Schieb et al. 1990). Figure 2.10 shows a photographic image of the Multicomponent Quartz Force Plate. Further information on the above four categories of slip resistance measuring devices are found in the recent literature (Brungraber 1977; Grönqvist et al. 1999; Chang et al. 2003b).

Pedestrian walkways and footpaths should be tested using a tribometer (floor slip resistance tester) to discover if there is a high propensity for slip and fall accidents on it, either dry and/or when polluted with water or lubricated with other contaminants such as kitchen grease, hydraulic oil, etc. As discussed in the above, there are numerous slip resistance testing apparatuses either in a laboratory (before or after installation) or on floors in situ produced around the world to measure both

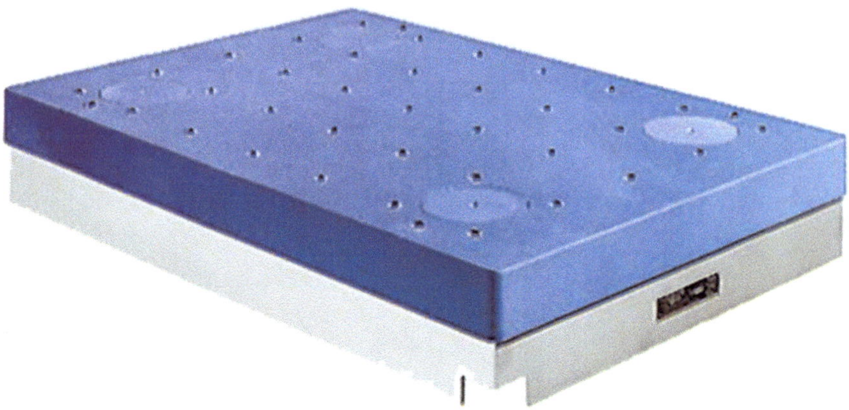

Fig. 2.10 A photographic image for the multicomponent quartz force plate (FP) manufactured by Kistler Instrument, Switzerland. Frictional and ground reaction forces are measured by the FP

static and dynamic COFs, but presently there are only a few that have been proven to be reliable for obtaining useful safety results and that have current official test methods. Table 2.1 summarizes laboratory based slip resistance tests with their usage and brief descriptions.

2.8.5 Slip Measuring Testers Used in This Book

Slip resistance measurements in this book were carried out by two pendulum-type dynamic friction testing machines that were designed and built to meet the International Standard Organization (ISO) requirements (ISO 1992; Stevenson 1997). Figures 2.11 and 2.12 show the dynamic friction testing machines developed by Stevensn et al. (1989) and Stevenson (1997), respectively.

As shown in Fig. 2.11, this testing machine consists of two hydraulic systems with an attached artificial foot, a force component transducer (Kistler 3-Component Dynamometer, Type 9257A), a desktop computer and an angular displacement transducer (Kim et al. 2013). This tester was designed to simulate the movement and loading of the foot during heel strike and initial slip and to quantitatively determine the slip requirement as DFC. The tester set up values could be adjusted to cover various parameters taken from biomechanical studies (human walking trials) such as heel contact angle, vertical load and its rate of increase and sliding speed.

To adjust the heel contact angle, tapered shims were inserted between the base of the pendulum and the last on which the shoe was mounted. The vertical load was set by the length of the pendulum adjusted by two nuts. Some fine adjustment to the vertical load was also possible by a pressure control on the oil fed to the vertical hydraulic cylinder. The sliding speed was set by a flow control valve.

Table 2.1 A summary of the laboratory based slip resistance tests

Floor slip resistance lab test	COF[a] assess	Procedure	Statement
ANSI[b] A137.1 (aka 'AcuTest'[c])	DCOF[d]	Specified in 2012 International Building Code for indoor wet areas	Like ANSI B101.3, but needs a higher minimum "passing" value and more stringent sensor prep
ANSI B101.1	SCOF[e]	Not a valid test for pedestrian safety, because it's a static test	Outdated by ANSI B101.3, a dynamic test
ANSI B101.3	DCOF	Using the BOT-3000E digital tribometer and for mostly indoors	Pendulum is more appropriate for outdoor surfaces due to its higher speed
ASTM C 1028-07	SCOF	Some specifiers still require it	An outdated and misleading test removed by ASTM in 2014
ASTM E 303 road skid test	DCOF	An American skid test for road surfaces	A simplified pendulum test with soft TRL rubber
Pendulum (CTIOA[f], HB 198/AS 4663-2013, or EN 13036-4)	DCOF	Dry and wet slip resistance appropriate to various conditions (pool deck, lobby, rest room, etc.)	The most widely used slip resistance tests worldwide. Endorsed by CTIOA since 2001
SlipAlert	DCOF	Qualitative demo mimics pendulum readings	A simple and durable machine is used that requires minimal training
Sustainable slip resistance pendulum	DCOF	Test whether slip resistance persists after significant abrasive wear by pedestrians	High pedestrian traffic areas must remain slip resistant
Tortus digital tribometer	DCOF	Dry and wet slip resistance up to high COF values with hard or soft rubber	Endorsed by CTIOA since 2001

[a]COF: coefficient of friction
[b]ANSI: American National Standards Institute
[c]DCOF AcuTest is an evaluation of the COF of a tile surface under known conditions using a standardized sensor prepared according to a specific protocol. Measurements are made with the BOT-3000, an automated and portable device that measures DCOF. The ANSI standard A137.1-2012 also allows the use of other equivalent tribometers
[d]DCOF: dynamic coefficient of friction
[e]SCOF: static coefficient of friction
[f]CTIOA: ceramic Tile Institute of America

In this tester, the test shoe was firmly attached to a last and mounted at the end of the pendulum mechanism. To minimise any movement, the shoe was nailed to the last. The floor surface specimen was glued to a steel plate that was bolted onto the force component transducer.

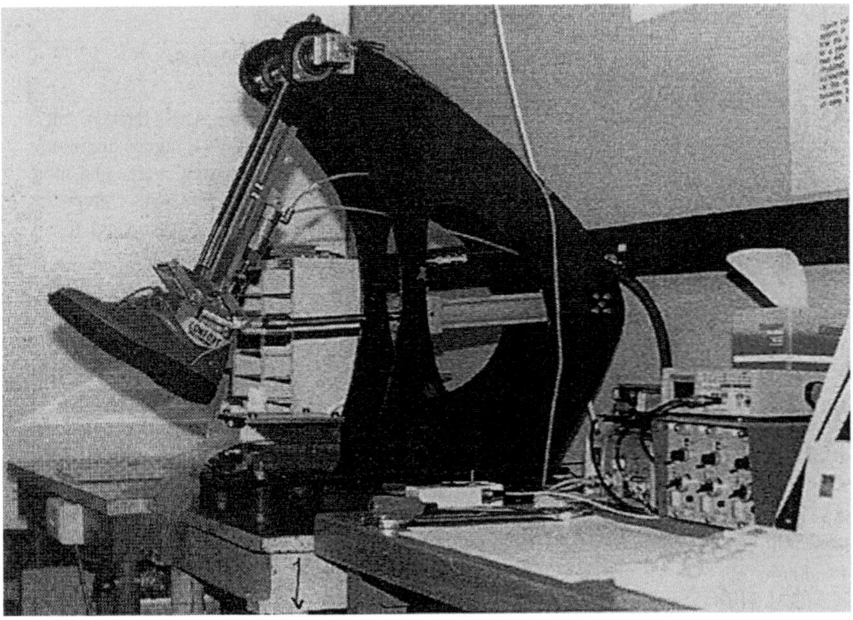

Fig. 2.11 A photographic image for a pendulum-type dynamic friction tester (Stevenson et al. 1989)

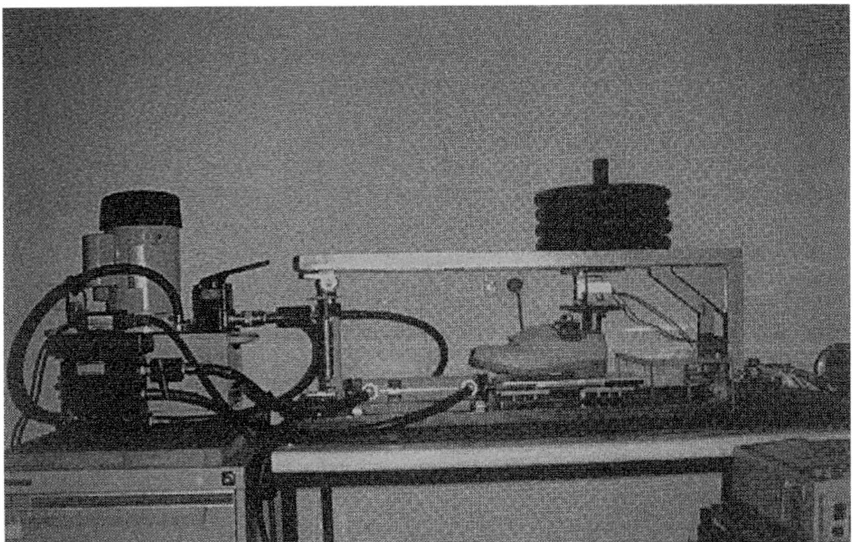

Fig. 2.12 A photographic image for a dynamic friction tester (Stevenson 1997)

In a test, the shoe was driven forward by the horizontal hydraulic cylinder to contact the floor sample surface at the heel edge. Another hydraulic cylinder was mounted at the end of the pendulum to simulate the body weight portion supported by the leading foot at heel strike. The two hydraulic cylinders were in a common circuit supplied by a pump which was driven by an electric motor.

As the shoe heel passed across the floor sample surface, the frictional (horizontal: ***H(F)***) and normal (vertical: ***V(F)***) components of the resultant force were measured by the force component transducer. The speed of the test was measured by a rotary potentiometer driven by the pendulum shaft. The force component signals and potentiometer voltage were recorded on a personal computer which continually calculated the ***H*/*V*** force ratio. During the tests, the normal force was kept around 350 N and the sliding speed was controlled at 40 cm/s based on gait studies (Redfern and Bidanda 1994; Jones et al. 1995). A heel contact angle of 9° was chosen by the result of previous biomechanical studies (Hoang et al. 1985, 1987).

Figure 2.12 presents the basic components of the test rig. It consists of two hydraulic systems to which the artificial foot is attached. The test shoe was filled with a polyester resin permitting a rigid connection to the machine. Then, the shoe heel is mounted to a last bolted to the base of a pair of load cells which monitor the vertical force on the shoe heel. A wedge between the shoe and the load cell is used to set the shoe heel angle.

The vertical load of 50 kg was applied partly by a share of the weight of the frame itself, but mainly by a set of weights resulting on top of the frame. The floor sample was screwed to a carriage which could move horizontally on low friction linear bearings. The movement of this carriage was controlled by a horizontal hydraulic cylinder. A vertical hydraulic cylinder was used to raise one end of the frame, thus lifting the shoe off the test surface. In order to measure the exact position from the shoe heel strike to heel off against the floor surface, a potentiometer was newly installed near the horizontal hydraulic cylinder.

In a test, the pressure of the vertical cylinder was released, allowing it to lower at a rate controlled by a flow valve. When the vertical force was fully applied, the horizontal cylinder was actuated, causing the carriage with the floor specimen to move at a controlled speed of 25 cm/s as the DCOF was measured. Other variables which were controlled in the tests were the heel contact angle (9°) and the rate of vertical loading on heel strike (5 kN/s). These values gave a reasonable simulation of heel strike conditions. The vertical and horizontal load cells were calibrated by dead weights of known mass.

2.8.6 Comparisons of Slip Measuring Devices

The versatilities of various devices and apparatus for assessing slip resistance properties have been compared and commented by several studies (National Research Council 1961; Bring 1964, 1982; Brungraber 1976; Braun and Brungraber 1978; Cohen and Compton 1982; Andres and Chaffin 1985; English 1990; Grönqvist et al. 1989, 1999).

Despite many years of investigations and fabrication of numerous testing machines, none of them is internationally adopted as a standard model yet, because each of them has some advantages and disadvantages in their designs and testing performance (Kim 2006b). In fact, because each testing instrument has different concepts, systemic parameters, and mechanical principles, it seems unreasonable to adopt a reference value without any citation to the instrument used for the slip resistance measurements (Kim 2006b; Kim and Smith 2003).

In the case of pendulum type devices, they require frequent calibrations and record only dynamic friction quantities which are not particularly critical in the initiation of a slip. The James Machine can only be used in a controlled laboratory condition for testing floor materials, not the actual floor surfaces due to the machine's heavy weight and bulkiness. On the other hand, the NBS—Brungraber Tester is easier to handle and can be considered as a portable James Machine. Tortus provides a permanent trace record when connected to a chart recorder, whereas most devices offer only a visual recording of the peak COF via an analogue meter or gauge.

The majority of slip meters require a number of replications in order to obtain a desirable level of statistical confidence in the measured quantities of COF and in measuring less than ideal surface conditions. This means that slip measurements need to be repeated as the floor surface is not being horizontal (inclined or ramp) or uniform status over the complete section rubbed by the shoe sole materials. For example, the small size of shoe samples on the UFTM, HPS, and British TRRL Pendulum Tester tends to interlock with less than uniform surfaces such as textured tiles or carpeted surfaces.

Additional problems include lack of movabilities and calibrations. The use of some other devices is also limited in certain locations such as steps due to the requirements of space on the horizontal plane for slip resistance measurements.

It seems to be more difficult to choose specific test conditions such as a test speed, heel strike angle (contact angle), and normal force. In the case of the test speed, it is controversial to choose an appropriate slip speed at which to drive the shoe heel across the floor surface. Normal walking speed varies from 1 to 2 m/s, but the forward speed of the shoe heel edge seems to be less than this speed just before the heel strike.

Strandberg (1983) reported that walking speed was varied from 0.06 to 0.32 m/s in his experimental results with different subjects. After sliding started, however, the shoe heel accelerated to a value above the walking speed. He concluded that the speed of a dynamic test should be in the range between 0 and 0.5 m/s. Perkins and Wilson (1983) also concluded from their experiments that

> Probably the ideal speed for high-speed measurements is 0.5–1.0 m/s since the foot and shoe can be travelling at this speed when the shoe heel tip touches the floor surface. Even if slip starts from a static situation, such is the acceleration that the speed of slip is about 0.5 m/s after only 0.01 m slip distance.

The strike angle at which the shoe heel first contacts the floor surface is an important factor in slipping (Kim and Smith 2003). This contact angle has proved to be an important variable because of its influence on the nominal contact area between the shoe heel and floor surface (Kim 2002a, b). Specifically, the contact area between the shoe heel and floor surface seems to play a dominant role in the sliding process when walking under lubricated environmental conditions (Moore 1972; Kim 2003a, b). Hence, observing the heel strike angle at a "correct angle" seems to be one of the most vital parts for the shoe heel during walking.

Hoang et al. (1987) conducted biomechanical studies to measure foot angles of 32 male subjects (height ranges: 155–180 cm) walking on a force platform. They found that the heel strike angle varied with walking speeds, step sizes, and subject's heights. On a horizontal surface, the strike angle laid with the range of 6°–10° measured from the floor surface.

In the case of the normal force, Strandberg and Lanshammar (1981) found that the leading foot was born up to 60% of the body weights, acting at the heel edge during the shoe heel contacted the floor surface.

Most of the testing devices for slip resistance are, however, not mechanically capable of simulating human gait parameters which are considered to crucial during actual slippage. Such gait parameters include horizontal and vertical shoe motions, normal and frictional forces, the normal force applying time, and sliding velocity. As indicated by Strandberg back in 1983, an ideal device should meet all the required demands and be reasonable to provide users' perspective in terms of cost/benefit ratios (Strandberg 1983).

Because such a device is obviously non-exist until at least now, Andres and Chaffin (1985) suggested that selecting a proper tester or technique would be a viable starting point for measuring slip resistance properties. They recommended that users should decide which characteristics of slip resistance needed to be investigated before the most appropriate technique(s) could be selected.

They also reminded that the requirement of static and dynamic COFs was only one aspect of a global assessment of slip and fall hazards. Therefore, extra efforts are urgently required to measure realistic COF readings that represent actual kinematics during critical gait phases when foot slippage starts (Grönqvist et al. 1993) and develop new concepts and theory models that allow to measuring, analysing, interpreting, and predicting slip resistance properties under a range of walking environments (Kim 2015a, b, 2016a).

2.9 Testing Standards and Safety Criteria for Slip Resistance Performance

2.9.1 Slip Resistance Test Methods and Safety Criteria

The essentials to assessing slip-related fall safety need an accurately measuring slip resistance test method, and a minimum numerical safety criterion such as 0.42, 0.50, and 0.60 to apply for the slip test results. It is generally recognised that there is no single device and standard for reproducing and quantifying the slipperiness of a floor surface yet. However, five methods currently co-operate internationally.

2.9.1.1 The Pendulum Slip Resistance Tester

The ASTM E303-93 (US), BS EN 13036-4:2011 (UK), and AS 4663:2013 and AS 4586:2013 (Australia) define the pendulum tester that is now a national standard for pedestrian slip resistance in 49 nations on five continents and has been endorsed by the CTIOA since 2001 (Sotter 2000; CTIOA 2001a; ASTM 2008; BSI 2011). It is the most widely used pedestrian slip resistance test method worldwide.

The pendulum uses a standardized piece of rubber (Four S rubber also known as Slider 96), which is set up to travel across the flooring sample for 123–125 mm, mounted onto the pendulum foot (see, Fig. 2.13). When the arm of the pendulum is set up to miss the flooring completely, the arm swings up to parallel from where it started, and the pointer (brought along by the arm holding the rubber slider) reads zero. Slippery flooring produces readings close to zero, and flooring which shows higher resistance to slipping give results further from zero—high numbers (such as those 36 and above) indicating suitable slip resistant flooring (UKSRG 2016).

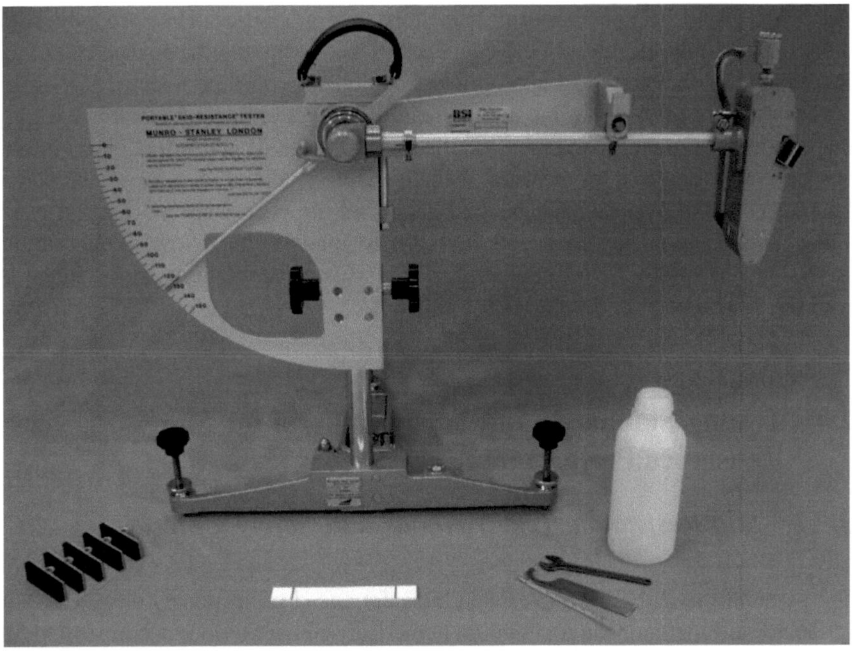

Fig. 2.13 A photographic image of a Munro British Pendulum Tester

Since 1971, the UK has established slip resistance standards based on the pendulum slip resistance tester. This test was developed for pedestrian traction by the US National Bureau of Standards in the 1940s and further refined in the UK (Giles et al. 1964). It was validated for pedestrian traction in 1971, together with its safety standards, in the UK over a period of 25 years by testing 3,500 real-world public walking areas (GLC 1971, 1985).

The tester has adopted as an ASTM standard (E 303) with slightly modifications for pedestrian traction measurement. The usual safety standard for a level floor is a minimum Pendulum Test Value (PTV) of 36. The pendulum is also the instrument used in the Sustainable Slip Resistance (SSR) test method, which measures the possible impact of years of use on potential flooring's slip resistance performance. The pendulum is also used elsewhere for determining the slipperiness of roads and airport runways.

Standards Australia HB 197:1999 as well as Standards Australia HB 198:2014 give detailed recommendations and guidelines of minimum wet Pendulum Test Slip Resistance Values (PTSRVs) for many different situations, such as external ramps, external walkways, pedestrian crossings, shopping center food courts, and elevator lobbies (Standards Australia 1999). There are also barefoot area recommendations based on the pendulum tests with a soft rubber slider (TRL rubber and also known as Slider 55). The Australian recommendations are presently the world's most detailed standards for the pedestrian wet slip resistance.

2.9.1.2 Tortus Digital Tribometer

The Tortus digital tribometer slip resistance test method is based on a proprietary or patented device, which is produced in the UK, and is in a category of slip resistance tester devices known as "drag-sled meters". Figure 2.14 shows a photographic image for the Tortus device sitting on a tile floor.

It travels across flooring under its own power at a constant speed with a piece of standardized rubber dragging on the flooring (Munro Instruments 2013). The amount of friction created by this piece of rubber as it is dragged across the flooring (dry or wet) is recorded and calculated by the machine as it travels a predetermined path length. An average number of DCOF is calculated by the machine after its run across the flooring has been completed. This is recorded as the "DCOF", or the amount of friction necessary to drag the standardized rubber across the flooring. High numbers (>0.50) indicate that it is difficult for the machine to drag the rubber across the flooring because it is anti-slip. Whilst low numbers (<0.50) means that the rubber easily slides across the flooring and is therefore slippery.

The Tortus becomes a primary instrument for assessing dry slip resistance in the latest Australian slip test standard—AS4586-2013 (SAI Global 2013). It also has been endorsed as a secondary standard by the CTIOA since 2001, with the pendulum being the primary standard. As compared to the pendulum tester, the Tortus tribometer can perform many slip tests in a short period of time under dry and wet environments, using both hard and soft rubbers. Unlike some tribometers, test

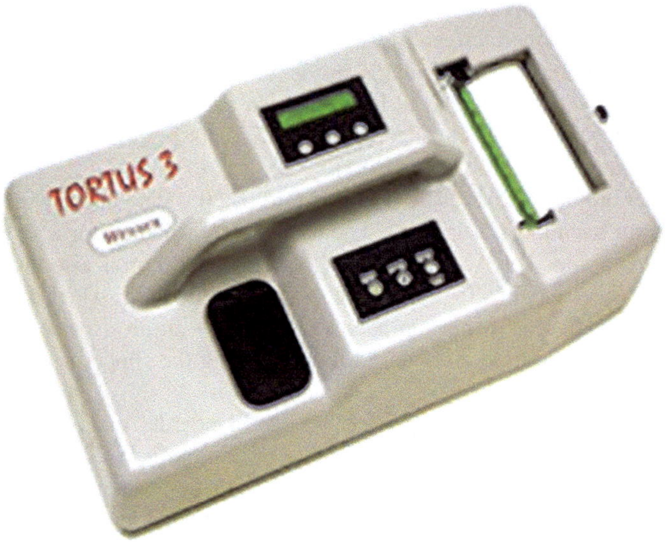

Fig. 2.14 A photographic image for the Tortus floor slip resistance tester

results from the Tortus tribometer are less influenced by an operator because an electronic button is pushed and then the test is run without any further help from the operator. The CTIOA has endorsed a minimum DCOF for level floors of 0.50 using the Tortus slip resistance test method (CTIOA 2001a).

2.9.1.3 BOT 3000 Slip Tester

The BOT 3000 is manufactured by Regan Scientific Instruments, Southlake, TX (Regan Scientific Instruments 2015). Figure 2.15 shows a photographic image of the BOT slip tester. It is self-propelled at 20 cm/s (7.9 in/s) and measures the SCOF and the DCOF against both wet and dry surface conditions with very little operator input. It is similar in operation to the Tortus tribometer except that the test pad is not spring loaded and applies a normal force of 21.3 N (4.8 lbs.) only after an operator activates the test which tends to lessen the effects of sticktion.

The machine pushes itself along the floor and the horizontal force applied to the pad is digitally measured and logged by the means of strain gauges. As the machine moves, the COF is calculated internally and both static and dynamic COFs are logged and displayed in real-time. Since this data is stored in onboard memory, it can be graphed and the minimum and maximum readings are easily available.

The BOT 3000 is the only machine approved and recognized to perform material testing according to ANSI B101.1 "Test Method for Measuring Wet SCOF of Common Hard-Surfaced Floor Materials Standard" and ANSI B101.3 "Test Method for Measuring Dry SCOF of Common Hard-Surfaced Floor Materials

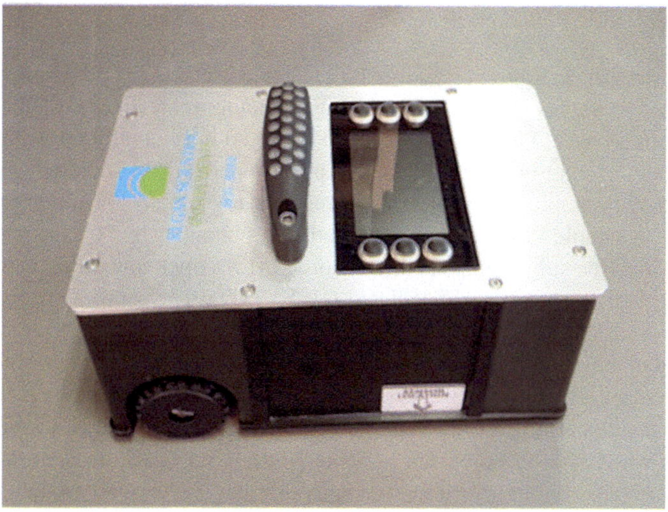

Fig. 2.15 A photographic image for the BOT 3000 slip tester

Standard" (American National Standards Institute (ANSI) 2012). The BOT 3000 is the only device approved for use by the Tile Council of North America (TCNA) for assessing and measuring the COF of tile (ANSI 2009). The TCNA had been using ASTM C1028 "Standard Test Method for Determining the Static Coefficient of Friction of Ceramic Tile and Other Like Surfaces by the Horizontal Dynamometer Pull-Meter Method" as the method for measuring the SCOF of tiles. In April 2012, an updated version of ANSI A137.1 "Specifications for Ceramic Tile" incorporates a DCOF testing method using the BOT3000.

The safety standards ANSI specifies for a level floor using the B101.3 dynamic test method is that a floor having a minimum DCOF of 0.43 has "high slip resistance". If the DCOF falls within the 0.30–0.42 range, then the test method states that the flooring is "acceptable, with an increased probability of slipping," and values below 0.30 are categorized as "low slip resistance" and a "higher probability of slipping" (ANSI 2012). The standard also states a minimum DCOF of 0.46 for ramps.

The BOT-3000 slip tester has passed the ASTM F2508-13 standard published in 2013 called the "Standard Practice for Validation, Calibration, and Certification of Walkway Tribometers Using Reference Surfaces" (ASTM 2013). Although this ASTM test method was created with good intentions, the test is often conducted and interpreted by operators who have conferred interests in their tribometer having passed the test. This means that a tribometer having passed this test should not be solely relied upon deciding whether a certain tribometer is trustworthy or not. A reliable tribometer also should be able to provide a reasonable precision statement as required by every test standard-publishing agency, such as the ASTM, ANSI, Standards Australia, and British Standards.

2.9.1.4 Variable Angle Ramp

The Variable Angle Ramp is a German developed method for obtaining pedestrian slip resistance values. Figure 2.16 shows a photographic image of the Variable Angle Ramp system. Flooring samples are mounted horizontally on the ramp tester and an operator clad in safety boots or bare feet performs a standardized walk up and down the sample while wearing a harness to stop the operator from being injured. The sample is slowly inclined until the operator slips on the surface. The angle at which the subject slips is then recorded. Two operators repeat this test three times and then an average is calculated (HSE 2009). The repeatability of this test method was extensively documented (Jung and Schenk 1988).

Tests can be performed under dry, wet with soapy water, and wet with oil environments. Over 150 safety criteria have been adopted in Germany and Australia for specific situations such as swimming pool decks, commercial kitchens, and restrooms which are based on the variable-angle ramp. But, the ramp itself is not readily portable, so this instrument is only for lab testing and is therefore quite limited in its effectiveness (Sotter 2000; CTIOA 2001b).

This signifies that you can't test your specific floor tiles using this method without ripping part of your floor up and putting it into the ramp tester. However,

Fig. 2.16 A photographic image for the variable angle ramp

since it is measuring human ambulation, it is considered by many to be the most realistic test method in existence, and the results of pendulum and drag-sled meter tests are sometimes compared with results from variable angle ramp tests to see if the results have a strong correlation. A good correlation with ramp test results can help a slip resistance test device become more widely used and accepted.

2.9.1.5 SlipAlert Slip Tester

SlipAlert is a new British slip tester. Figure 2.17 shows a photographic image of the SlipAlert slip tester. It is a roller-coaster type tribometer that is designed to mimic the readings of the modern pendulum. It measures the slipperiness of a floor in order to assess whether it is safe. SlipAlert has been specifically designed to simulate a real slip and to correlate with the TRL Pendulum. It has been used for field testing, but is of limited utility in laboratory testing because it requires a long path length of flooring to conduct slip tests (SlipAlert 2011). As the SlipAlert is an exclusive device, it does not have an official American standard test method, but it now has an official British standard for its use in the field: BS 8204-6:2008 (Standards Centre 2010).

The SlipAlert "car" has a rubber slider on its bottom, which slides across the flooring after running down a fixed ramp. If the SlipAlert stops short, then the flooring is slip resistant, but if it slides a long distance then the floor is considered slippery. There is a digital readout on the device that records the maximum distance the SlipAlert has traveled across the flooring, and a safety criterion graph which

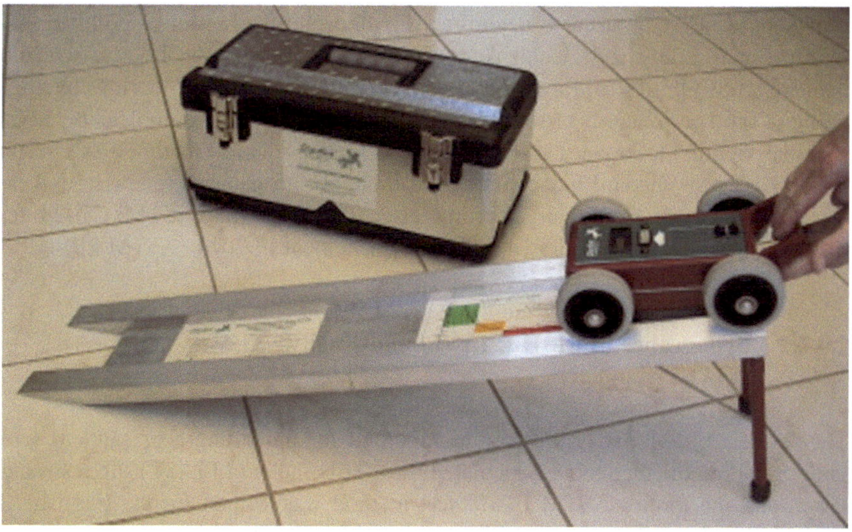

Fig. 2.17 A photographic image for the SlipAlert slip tester

interprets the results. Tests can be done under dry and wet environments, and extensive research by U.K. and Australian government agencies have resulted in several endorsements of this test device for in situ testing.

2.9.2 *Undependable Test Methods and Removed Standards*

2.9.2.1 ASTM C1028-07

This test method covers the measurement of static coefficient of friction (SCOF) of ceramic tile or other surfaces under both wet and dry conditions whilst utilizing Neolite heel assemblies. This test method can be used in the laboratory or in the field so often has operated in the US for assessing walkway safety (ASTM 2014).

Formerly under the jurisdiction of Committee C21 on Ceramic Whitewares and Related Products, however, this test method was withdrawn in February 2014. This standard is being withdrawn without replacement due to its limited use by industry (ASTM 2014). Safety criteria based solely on the SCOF are too often misleading where floors get wet or otherwise lubricated in use (CTIOA 2001b; ASTM 2005; Astrachan 2007; Powers et al. 2007). For example, static friction tests can be conducted before and after a fresh coat of wax is applied to a floor surface to validate whether the SCOF has not changed noticeably with a new wax, or the SCOF readings can be taken at different times during a day to make sure that any dirt and dust is not making the slip resistance condition of floorings worse throughout the day.

Monitoring changes in dry SCOFs can be a useful practice. However, SCOF tests should never be used to determine if a floor surface is wet or polluted. The Occupational Safety and Health Administration (OSHA) in the USA has recommended an SCOF of 0.50 for workplace environments, but often floorings having a rating of 0.60 or greater is proven by reliable test devices (and multiple slip and fall accident victims) to be very slippery when wet using this test method. As a result, this ASTM test standard has expired, and there are no plans to renew it as it has caused more confusion than anything else. Therefore, the ASTM C1028 test method was officially withdrawn with no replacement in 2014 (ASTM 2014).

2.9.2.2 Brungraber Mark II (PIAST) and English XL (VIT) Tribometers

The former ASTM F1677 was the test method for the Brungraber Mark II (also known as Portable Inclinable Articulated Strut Tribometer or PIAST) slip resistance tester, and the former ASTM F1679 test method was written for the English XL Variable Incidence Tribometer (VIT), respectively. Shortly after being published in 2004, ASTM withdrew these two standards in 2006, with no replacement.

Regarding ASTM 1677, ASTM provides the following similar "Withdrawn Rationale":

> Formerly under the jurisdiction of Committee F13 on Pedestrian/Walkway Safety and Footwear, this test method was withdrawn as an active ASTM standard by action of the Committee of Standards (COS) on September 30, 2006 for failure to include an approved precision statement (violating Section A21 of the Form and Style for ASTM Standards), and for including reference to proprietary apparatus where alternatives exist (violating Section 15 of the Regulations Governing ASTM Technical Committees) (ASTM 2006a).

ASTM also offers the following similar "Withdrawn Rationale" for F1679:

> Formerly under the jurisdiction of Committee F13 on Pedestrian/Walkway Safety and Footwear, this test method was withdrawn as an active ASTM standard by action of the Committee of Standards (COS) on September 30, 2006 for failure to include an approved precision statement (violating Section A21 of the Form and Style for ASTM Standards), and for including reference to proprietary apparatus where alternatives exist (violating Section 15 of the Regulations Governing ASTM Technical Committees) (ASTM 2006b).

Using these two instruments, different test labs showed very different answers on identical tiles amongst interlaboratory studies. These findings suggested that both test methods (or tribometers) were unreliable and unable to provide "reasonable precision statements" for slip resistance evaluations. However, the English XL Variable Incidence Tribometer (VIT) is still used commonly in the US and worldwide, mostly by professional experts, who have discovered that the test results can be considerably influenced by how the actuating button is pressed by the user.

2.9.3 Clean and Dry and Wet Slip Resistance Measurements

2.9.3.1 Importance of Dry Slip Resistance Measurement

Dry slip resistance measurements can provide valuable information with or without wet testing. There are several reasons for the importance of dry slip resistance measurements:

(1) Clearly, the laws for wet (or lubricated) surfaces are different to those for dry friction, because of the effect of lubricant films between the shoe heel and floor surface. In the case of the shoe-floor traction system, the types of contact surfaces are largely changed by different walking environment such as dry, dusty, fibrous, granular, organic and wet, moist or contaminated. As a result, the friction force for the shoe outsoles and heels moving over walkway surfaces increases with increase in contact area. This aspect is in violation of the 'classical' laws of friction (Bonstingl et al. 1975). However, for the dry friction, the friction force is independent of the contact area (Jenson and Chenoweth 1990), which is contrary to the above findings. This aspect indicates that dry friction behaviours attribute to a complex molecular-mechanical interaction that occurs between the contacting surfaces of shoe heel and floor surface. This

complex mechanism of the contact interface between them is mainly caused by the combined effects of asperity deformation, ploughing by hard surface asperities and wear particles, and adhesion between flat surfaces (Suh and Sin 1981; Czichos 1986; Kim 2015a, c, 2016a, b). As a result, wear developments of the floor surfaces seem to be occurred and can substantially affect slip resistance properties (Leclercq and Saulnier 2002; Kim and Nagata 2008b; Kim et al. 2013; Kim and Alduhishy 2013; Kim 2015c, d). Even though the significance of flooring wear issue, understandings of flooring wear behaviours, accompanying tribo-physical mechanisms, and their resultant consequences against slip resistance performance have remained as an unexplored area. Therefore, analytical investigations for the dry sliding friction behaviours should be thoroughly investigated and analysed for slip resistance measurements.

(2) Dry slip resistance measurements followed immediately by wet (or lubricated) tests on the same floor can help to demonstrate that the slip resistance test method is capable of measuring both high and low values on the targeted flooring.
(3) A floor surface may be dry and/or appear clean, but be slippery due to a thin film of contaminant, for instance grease in or near a restaurant kitchen or parking structure.
(4) Dry environment data can also help to diagnose problems such as furniture polish overspray, inadequate maintenance, airborne cooking fat or oil that settles to the floor overnight.
(5) A rare small spill on a normally dry floor may have occurred too after cleaning.
(6) Many claims of slips and falls are made that involve dry floors.

The actual causes of the slip and fall incidence may also include footwear, substance abuse, illness, or many other factors unrelated to the floor surface itself. Dry slip resistance testing—particularly before the alleged accident and on a periodic monitoring basis (such as traction auditing)—can help to establish that the floor is safe (or anti-slip) when dry.

2.9.3.2 The Sine Qua Non for Slip Safety Assessments

Although a number of different types of slip resistance measuring devices and testers have been developed, there is no universally accepted test apparatus or measuring method. The recent literature has shown wide disagreements on the results of these devices, even using the same test conditions and environments (Harris and Shaw 1988; McElvaney 1993; Dutruel and Degas 1997; Grönqvist et al. 1998, 1999; Ricotti et al. 2009). These discrepancies may stem from design principles, kinematic and kinetic characteristics of measuring devices, mechanical and physical properties of sliding materials, and test and operation conditions.

The US OSHA has long recommended a minimum COF quantity of 0.50 for the workplace safety without any detailing information such as how the COF is to be determined. Because different slip resistance test methods and measuring devices produce different results, particularly under wet environmental conditions, this makes a great deal of confusions and mistakes for objective assessments of fall safety. As a result, the same floor could show quite different COF readings such as 0.4, 0.5, 0.6, or 0.7 (or anywhere in between) depending entirely up on what test method and tool is used. The OSHA recommendation is, therefore, meaningless, but has caused misperceptions for many years. However, the following three slip testers are frequently adopted as valid ones for the appropriate purposes (Sotter Engineering Corp. 2017):

(1) BOT-3000E slip tester:

This tester is cost-effective, rapid and automatically documented dry and wet testing, both static and dynamic, with various available test foot materials.

(2) Pendulum Slip Resistance Tester:

This tester is for dynamic COF testing under wet and dry environments with hard and/or soft rubber sliders. This is for testing outdoor and barefoot areas. This tester provides sustainable slip resistance testing results and application of long-established detailed situation-specific standards for swimming pool deck and outdoor ramp.

(3) Tortus Tribometer:

This tester is for testing where pedestrians are not likely to be running, and where numerous wet and dry tests are needed with hard and/or soft rubber sliders.

As discussed in the above, however, none of these slip resistance testers is capable of doing all the tasks for measuring slip resistance performance. Above all, overall performance of the above three slip testers is significantly affected by many factors such as floor material types, texture characteristics, and presence of contaminants on the floor surface.

In summary, although slip resistance measurement results from different testers may not be necessarily compatible with each other against any targeted flooring material and encountered environmental condition, it needs to be advised that comparing the measured slip resistance data from a field investigation with different types of slip meters can be ineffective. Therefore, to evaluate slipperiness in the real-world walking situations, the performance of slip meters operated still require a great deal of research challenges to improve their accuracy, reliability, validity.

2.10 Relationships Between Human Gait and Slip Resistance Properties

The issues of slip resistance related to human gait were extensively investigated in the literature (Leamon and Li 1990; Myung et al. 1992a, b; Burnfield and Powers 2006; Powers et al. 2007). Leamon and Li (1990) demonstrated that there was usually a sliding between the shoes and floor surfaces, immediately followed heel strikes. If human subjects regain control, then sliding was stopped and subjects regained their balance. The sliding distance can change from an undetected micro-slip (<3 mm) that requires no corrective actions to an uncontrollable slide (>10 cm) that eventually results in a fall if the relative velocity or sliding distance between the shoe heel and floor surface exceeds 50 cm/s or 10 cm, respectively.

Myung et al. (1992a, b) also reported that the slip distance immediately following heel strikes was inversely proportional to the static COF (SCOF) of the floor surface, although their results were not statistically significant. However, a high level of friction could be beneficial to reduce the slip distance so it seemed to improve slip resistance.

Burnfield and Powers (2006) investigated the relationship between measures of floor surface slip resistance and an individual's peak utilised coefficient of friction (UCOF) on the probability of a slip occurring during level walking. Powers et al. (2007) assessed the viability of using slip risks (as quantified during human subject walking trials) to create a reference standard against which tribometer readings could be compared. Their study reported that only two of the nine tribometers tested (Tortus II and Mark III) met our compliance criteria by both correctly ranking all six conditions and differentiating between surfaces of differing degrees of slipperiness. This finding reinforces the need for developing objective criteria to ascertain which tribometers effectively evaluate floor slipperiness and pedestrians' risks from slipping and falling.

Whether static or dynamic friction is more relevant to the contact-sliding mechanism between the shoe and floor is still debatable. Ekkubus and Killey (1973) suggested that static friction was more applicable than dynamic friction because the foot (or the shoe heel) was static relative to the floor surface when walking. They also suggested that dynamic friction would be valid only after the foot started to slip so could not be related to normal walking conditions. However, Lanshammar and Strandberg (1983) showed that the horizontal velocity of the heel edge relative to the floor surface had small but "non-zero velocities" at the heel strikes in contrast to the assumption of "zero velocity" by the majority of studies.

Tisserand (1985) investigated the significance of static versus dynamic friction by an experiment in which subjects wore shoes of different material types on either foot. He reported that the subjective ranking of the slipperiness of particular shoes correlated highly with the measured DFCs. Based on this evidence, he concluded that dynamic friction is more applicable to simulate human walking than a static one.

Other studies also suggested that the DFC measurements for standards were more relevant to the biomechanics of slips and falls than SCOF (Perkins 1978;

Strandberg and Lanshammar 1981; Hoang et al. 1987; Redfern and Bidanda 1994). Another reason for favouring dynamic friction tests was that they seem to more realistically assess the effects of contaminants on the floor surface because this could not be duplicated in static friction tests (Hoang et al. 1987).

Perkins (1978) and Strandberg and Lanshammar (1983) introduced an $\frac{H}{V}$ (horizontal force/vertical force) diagram to study the characteristics of walking and slipping. The diagram trace suggested by Perkins and Wison (1983) is reproduced in Fig. 2.18.

The Point 1 in Fig. 2.18 indicates a forward force on the initial heel contact that must be resisted by friction if the step is to remain under control. The Point 2 shows a brief backwards force, whilst the Point 3 results from the further force in the forward direction requiring friction to prevent slipping. This means that the point at which slipping is most likely to occur is the Peak 3.

Perkins (1978) also analysed both horizontal and vertical components of forces exerted between the shoe and floor surface during normal walking. He drew the following conclusions:

(1) During normal walking, falls could result as the foot slid forward at the moment that the shoe heel contacted the floor surface. At this instant, the average value of the $\frac{H}{V}$ ratio was approximately 0.2 for eight subjects, and the maximum values were about 0.33 and 0.25, respectively.
(2) If the length of slip exceeded 10–15 cm, a loss of balance occurred.

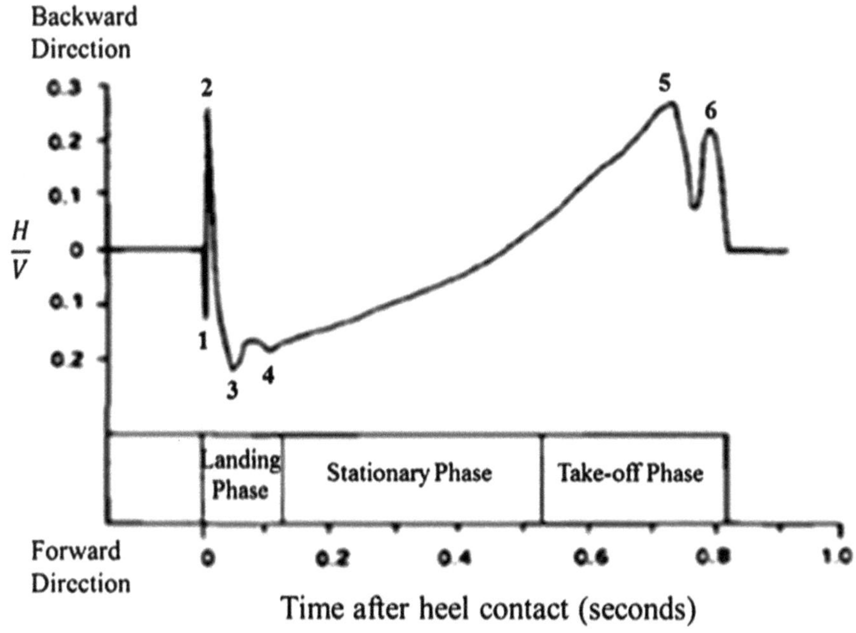

Fig. 2.18 The Perkins' H/V diagram for normal walking step (Perkins and Wilson 1983)

(3) In certain cases, the subject regained balance after slipping a short distance. This was explained by the fact that the entire shoe rapidly came into contact with the floor surface after the onset of slips and thus increased the level of friction.

The study results of Strandberg (1983) also showed a brief peak such as the Point 1 in Fig. 2.18. He further stated that "in most of the recorded experiments the shoe heel was sliding upon heel strike". Strandberg and Lanshammar (1981) conducted a biomechanical study with four subjects and concluded that

(1) A slipping of the shoe heel on the floor surface at the contact moment was observed during most of the tests. This movement was generally not noticed by the subject and occurred even if there was no lubricant on the floor surface.
(2) On average, the critical slipping movement occurred at 0.05 s after the shoe heel touched down. At this instant, the vertical force was equivalent to approximately 60% of the body weight and was applied to the sliding interface between the shoe heel (which made an approximately 5° angle) and the floor surface.
(3) Once initiated, a slip would progress if the $\frac{H}{V}$ ratio continued to increase after it begun.
(4) When the slip speed exceeded 0.5 m/s, or when the length of slip exceeded 10 cm, the slip generally led to falls.
(5) When there were no dangerous slips, the maximum value of the $\frac{H}{V}$ ratio at the beginning of the rolling action of the foot, was equal to 0.17 on average and was detected 90 ms. after the shoe heel contacted.
(6) When a slip did not lead to a fall, the $\frac{H}{V}$ ratio changed between 0.1 and 0.2 during the time that the feet sled forward.
(7) Sliding movements to the rear that occurred just before the toes were lifted from the floor surface never led to a loss of balance, or to a fall.

For the measurement of DFCs, Strandberg and Lanshammar (1981) suggested that slip resistance testers utilised certain variables from the crucial gait phase. Of many important parameters related to the crucial gait phase, Redfern and Bidanda (1994) classified the vertical force, shoe heel velocity, and shoe contact angle as the most significant biomechanical parameters. Shoe materials, floor types, and contaminants were classified as important environmental factors.

Strandberg and Lanshammar (1981) also mentioned that the vertical force was one of the particularly important variables that slip resistance meters should reproduce based on tribological and practical experiences. Grönqvist et al. (1989) and Wilson (1990) recommended that the range of vertical force in the testers varied from almost a whole-body weight to 50% of body weight. However, Chaffin et al. (1992) argued that although the basic assumption in computing the COF was that the friction force increased proportionally to the normal force holding the contacting surfaces together under some circumstances, this might not be true.

The effect of the normal load magnitude is not simple to predict. When a soft material such as crepe is loaded heavily on a rough surface, the effect of deformation forces on friction would be larger than when using a smooth surface or a harder shoe material (Chaffin et al. 1992). That is, an interaction mechanism between the shoe heel and floor surface exists so that this tribo-physical system makes to complicate generalising the interpretations and predictions of COF results under light and heavy loads. This fact is particularly troublesome for selecting a proper tester because some COF testing procedures use very light normal load (<10 N), but others advocate quite a high normal load (about 1000 N).

The other important issue to consider is to understand where the pedestrians place their foot to make contact with the walking surface. There are the requirements of adequate friction to resist slips (Perkins 1978; Strandberg and Lanshammar 1981) and the avoidance of excessive friction to cause trips (Grönqvist et al. 1989). This means that the use of roughened walking surface has a beneficial effect of raising the COFs significantly under both dry and lubricated surface conditions. However, the effect of walking on such high traction surfaces is problematic when walkers must continually rotate their feet.

If the shoe cannot be easily rotated in such floors, then additional torsional stress is placed on the knees and lower back areas. This "trade-off" effect between the surface roughness and walking comforts seems to be an important issue to investigate but it is scarce to find any detailed information explored in the current literature so definitely requires further studies (Kim et al. 2013). In this sense, the underfoot surfaces should be efficiently designed and built with optimal surface finishes that allow such topographic structures and comfort walking so that pedestrians would not require any adjustment to their gait.

The velocity of shoe heels at a striking moment is critically related to the occurrence of slips and falls (Strandberg and Lanshammar 1981; Perkins and Wilson 1983; Strandberg 1983; Rhoades and Miller 1988; Chambers et al. 2003; Lockhart et al. 2005).

Strandberg (1983) demonstrated that the shoe heel speed from different subjects varied from 0.06 to 1.7 m/s. He also found that the heel speed at skid started changed from 0.08 to 0.32 m/s. After skid started, the shoe heel accelerated to a value above that of walking speed. He accordingly concluded that the speed of a dynamic test should be in the range of 0–0.5 m/s.

Perkind and Wilson (Perkins and Wilson 1983) stated from their study that "probably the ideal speed for high-speed measurements is 0.5–1.0 m/s since the foot and shoe can be travelling at this speed when the shoe heel contacts the ground. Even if slip starts from a static situation, such is the acceleration that the speed of slip is about 0.5 m/s after only 0.01 m slip distance". Considering the slip speed issue as a controlling variable would add tremendous amounts of test volumes to assess the merits of shoe and floor materials. Therefore, it is necessary to identify a shoe test speed which represents real slip risks and keeps the speed during the planned tests.

The shoe strike angle is also an important factor in detecting any slip potential. Any change of the heel strike angle can affect the formation of a contact area between the shoe heel and floor surface so this result plays a dominant role in the sliding friction process under lubricated environments in particular (Moore 1972).

This means that the surface topographies of both shoes and floors also can be modified by the change of contact area and consequently slip resistance properties (Kim 1996a, b, c, d, 2004a, b, 2006a, b, 2015a, 2016a, b; Kim and Smith 1998a, b, 1999, 2000, 2001a, b, c, 2003; Kim et al. 2001, 2013; Kim and Nagata 2008a, b; Kim and Alduhishy 2013). Hence, striking at a "correct angle" seems to be one of the most important parts of the shoe in walking trials (Hoang et al. 1987).

Several studies suggested that the heel strike angle found to be around 5° (Strandberg and Lanshammar 1981; Grönqvist et al. 1989; Wilson 1990). However, Redfern and Bidanda (1994) reported that the heel strike angles between 5° and 15° were not practically significant.

By measuring COFs using different types of slip resistance testers, significant amounts of attempts have been made to develop a standard for the pedestrian safety. However, the earlier works clearly demonstrate the recognition of complex array of factors that affect the risks of slip and fall on walkway surfaces (Adler and Pierman 1979).

For example, Leclercq et al. (1995) reported that the measurement results of slip resistance were not identical to the new shoes with the same model. They stated that different manufacturing and storage conditions could explain those variations. Skiba et al. (1985) underlined the fact that the composition of shoe soles and the manufacturing process certainly influenced on slip resistance properties.

There seems to be growing consensus that friction measurement between the shoe heel/sole and floor surface is only one of many factors that relate to pedestrian fall incidence (Marosszeky 1985; Kim 1996a, b, c, d, 2004a, b, 2006a, b, 2015a, b, c, d, 2016a, b; Kim and Smith 1998a, b, 1999, 2000, 2001a, b, c, 2003; Kim et al. 2001, 2013; Kim and Nagata 2008a, b; Kim and Alduhishy 2013).

This finding clearly addresses that a single COF index or quantity can not accurately reflect slip resistance performance between the shoe and floor. Therefore, a thorough understanding of such complex mechanisms of slip resistance properties is urgently required to fully explore their multifaceted characteristics.

2.11 Chapter Summary

Pedestrian fall incidents resulting from slips or trips are major concerns. There have thus been prolonged efforts to understand the main causes of such incidence in order to reduce its severity and fatalities. However, pedestrian fall incidents are caused by multifactorial risk factors. These risk factors signify that fall safety assessment is a highly complex area to study, where the likelihood of a slip and fall is a function of a variety of elements such as surfaces, environmental conditions, and individual users.

Recognising that widening the perspectives may better define the role of multi-factors in detection and prevention of fall incidence, but this book is mainly focused on exploring environmental (floorings and shoes) correlations for slip and fall incidents under a range of environmental conditions as a starting point.

It has been found that the most common precipitating event leading to a fall is a loss traction or slip resistance between a shoe heel/sole and floor surface, followed by trips, misstep, loss of support, and postural overextension (Cohen and Compton 1982). Slip resistance property between the footwear and underfoot surface is of great importance for preventing slip and fall incidence and has been measured as a form of a COF. In this context, knowledge about friction demand and the friction available has been recognised as a main key factor to fall assessment.

A range of apparatuses and devices have been fabricated and developed to quantify frictional behaviours between the shoes and floors. Despite many years of investigations and constructions of numerous testing machines, none of them is internationally adopted as a standard model yet, because each of them has some advantages and disadvantages in their designs and testing performance (Kim 2006b). In fact, because each testing instrument has different concepts, operating parameters, and mechanical principles, it seems unreasonable to adopt a reference value without any citation to the instrument used for the slip resistance measurements (Kim 2006b).

There seems to be growing consensus that friction measurement between the shoe heel/sole and floor surface is only one of many factors that relate to pedestrian fall incidence (Marosszeky 1985; Kim 1996a, b, c, d, 2004a, b, 2006a, b, 2015a, b, c, d, 2016a, b; Kim and Smith 1998a, b, 1999, 2000, 2001a, b, c, 2003; Kim et al. 2001, 2013; Kim and Nagata 2008a, b; Kim and Alduhishy 2013). This finding clearly addresses that a single COF index or quantity cannot accurately reflect slip resistance performance between the shoe and floor. Therefore, thorough understandings of such complex mechanisms on slip resistance properties are urgently required to fully explore their multifaceted characteristics.

References

Abrahamsen, B., van Staa, T., Ariely, R., Olson, M., & Cooper, C. (2009). Excess mortality following hip fracture: A systematic epidemiological review. *Osteoporosis International, 20* (10), 1633–1650.

Agostini, J. V., Baker, D. I., & Bogardus, S. T., Jr. (2001). Prevention of falls in hospitalized and institutionalized older people. In K. G. Shojania, B. W. Duncan, & K. M. McDonald (Eds.), *Making health care safer: A critical analysis of patient safety practices* (No. 43). Rockville, MD: Agency for Healthcare Research and Quality; Evidence Reports/Technology Assessments).

Adler, S. C., & Pierman, B. C. (1979). *A history of walkway slip-resistance research at the National Bureau of Standards* (National Bureau of Standards Special Publication 565). U.S. Government Printing Office, Washington

American National Standards Institute (ANSI). (2009). *Letter of interpretation: ANSI/NFSI B101.1-2009*. National Floor and Safety Institute.

American National Standards Institute (ANSI). (2012). *B101.3 test method for measuring wet DCOF of common hard surface floor materials*. National Floor and Safety Institute. Available at: http://www.nfsi.org/ansi_b101_3.php

American Society for Testing and Materials (ASTM). (2005). *American society for testing and materials subcommittee 21.06: Meeting minutes 26*. May 3, Orlando, Florida USA.

American Society for Testing and Materials (ASTM). (2006a). *ASTM F1677-05 standard test method*. Withdrawn. Available at: http://www.astm.org/Standards/F1677.htm

American Society for Testing and Materials (ASTM). (2006b). *ASTM F1679-04e1 standard test method*. Withdrawn. Available at: http://www.astm.org/Standards/F1679.htm

American Society for Testing and Materials (ASTM). (2008). *ASTM E303-93*. West Conshohocken, PA, USA: ASTM International.

American Society for Testing and Materials (ASTM). (2013). *ASTM F2508-16: Standard practice for validation, calibration, and certification of walkway tribometers using reference surfaces*. West Conshohocken, PA, USA: ASTM International.

American Society for Testing and Materials (ASTM). (2014). *ASTM C1028-07e1: Standard test method for determining the static coefficient of friction of ceramic tile and other like surfaces by the horizontal dynamometer pull-meter method*. West Conshohocken, PA, USA: ASTM International.

Andres, R. O., & Chaffin, D. B. (1985). Ergonomic analysis of slip-resistance measurement devices. *Ergonomics, 28*(7), 1065–1080.

Andres, R. O., O'Connor, D., & Eng, T. (1992). A practical synthesis of biomechanical results to prevent slips and falls in the workplace. In S. Kumar (Ed.), *Advances in industrial ergonomics and safety* (pp. 1001–1008). IV: Taylor & Francis.

Ashton-Miller, J. A. (2005). Age-associated changes in the biomechanics of Gait and gait-related falls in older adults. In J. M. Hausdorff & N. B. Alexander (Eds.), *Gait disorders evaluation and management* (Chapter 4, pp. 63–100).

Astrachan, E. (2007). *Installer update: Updates to an American method for measuring coefficient of friction*. Tile Dealer, November & December

Baker, S. P., & Harvey, A. H. (1985). Fall injuries in the elderly. *Clinics in Geriatric Medicine, 1*(3), 501–509.

Bell, J. (1997). The best foot forward. *Occupational Health & Safety, 66*(12), 26–28.

Bentley, T. A. (2009). The role of latent and active failures in workplace slips, trips and falls: An information processing approach. *Applied Ergonomics, 40*(2), 175–180.

Berg, K. O., Wood-Dauphinee, S. L., Williams, J. I., & Maki, B. (1992). Measuring balance in the elderly: Validation of an instrument. *Canadian Journal of Public Health, 83*(Suppl 2), S7–S11.

Blake, A. J., Morgan, K., Bendall, M. J., Dallosso, H., Ebrahim, S. B., Arie, T. H., et al. (1988). Falls by elderly people at home: Prevalence and associated factors. *Age and Ageing, 17*(6), 365–372.

Blin, J., Baron, J. C., Dubois, B., Pillon, B., Cambon, H., Cambier, J., et al. (1990). Positron emission tomography study in progressive supranuclear palsy: Brain hypometabolic pattern and clinicometabolic correlations. *Archives of Neurology, 47*(7), 747–752.

Bonstingl, R. W., Morehouse, C. A., & Niebel, B. W. (1975). Torques developed by different types of shoes on various playing surfaces. *Medicine and Science in Sports, 7,* 127–131.

Boushon, B., Nielsen, G., Quigley, P., Rutherford, P., Taylor, J., Shannon, D., et al. (2012). *How-to guide: Reducing patient injuries from falls*. Cambridge, MA: Institute for Healthcare Improvement; 2012. Available from: www.IHI.org

Brady, R. A., Pavol, M. J., Owings, T. M., & Grabiner, M. D. (2000). Foot displacement but not velocity predicts the outcome of a slip induced in young subjects whilst walking. *Journal of Biomechanics, 33*(7), 803–808.

Braun, R., & Roemer, D. (1974). *Influence of waxes on the static and dynamic friction* (pp. 60–72). Soap/Cosmetics/Chemical Specialties.

Braun, R., & Brungraber, R. J. (1978). A comparison of two slip-resistance testers. In C. Anderson & J. Seene (Eds.), *Walkway surfaces: Measurement of slip resistance, ASTM STP 649* (pp. 49–59). American Society for Testing and Materials.

Bring, C. (1964). *Friction and slipping*. Stockholm, Sweden: The National Swedish Council for Building Research.

Bring, C. (1982). *"Testing of slipperiness", document D5*. Stockholm, Sweden: The National Swedish Council for Building Research.

British Standards Institution (BSI). (2011). *Road and airfield surface characteristics. Test methods. Method for measurement of slip/skid resistance of a surface. The pendulum test.* London, England: BSI.

Brungraber, R. J. (1976). *An overview of floor slip-resistance research with annotated bibliography* (National Bureau of Standard, NBS Technical Note 895). Washington: U.S. Department of Commerce, National Technical Information Service, PB-248 985.

Brungraber, R. J. (1977). *A new portable tester for the evaluation of the slip-resistance of walkway surfaces* (National Bureau of Standards Technical Note 953). Washington: U.S. Department of Commerce/National Bureau of Standards.

Bunterngchit, Y., Lockhart, T. E., Woldstad, J. C., & Smith, J. L. (2000). Age related effects of transitional floor surfaces and obstruction of view on gait characteristics related to slips and falls. *International Journal of Industrial Ergonomics, 25*(3), 223–232.

Burdett, R. G., Borello-France, D., Blatchly, C., & Potter, C. (1988). Gait comparison of subjects with Hemiplegia walking unbraced, with ankle-foot orthosis, and with air-stirrup(R) brace. *Physical Therapy, 68*(8), 1197–1203.

Burnfield, J. M., & Powers, C. M. (2006). Prediction of slips: An evaluation of utilised coefficient of friction and available slip resistance. *Ergonomics, 49*(10), 982–995.

Carroll, D. L., Dykes, P. C., & Hurley, A. C. (2010). Patients' perspectives of falling whilst in an acute care hospital and suggestions for prevention. *Applied Nursing Research, 23*(4), 238–241.

Ceramic Tile Institute of America (CTIOA). (2001a). *Floor safety reports: No. 1, portable methods*. Ceramic Tile Institute of America. Available at: http://www.ctioa.org

Ceramic Tile Institute of America (CTIOA). (2001b). *Floor safety reports: No. 3, endorsement of improved test methods and slip prevention standards for new flooring*. Ceramic Tile Institute of America. Available at: http://www.ctioa.org

Chaffin, D. B., Woldstad, J. C., & Trujillo, A. (1992). Floor/shoe slip resistance measurement. *Journal of American Industrial Hygiene Association, 53*(5), 283–289.

Cham, R., & Redfern, M. S. (2002). Changes in gait when anticipating slippery floors. *Gait & Posture, 15*(2), 159–171.

Chambers, A. J., Margerum, S., Redfern, M. S., & Cham, R. (2003). Kinematics of the foot during slips. *Occupational Ergonomics, 3*(4), 225–234.

Chang, W. R., Grönqvist, R., Leclercq, S., Brungraber, R. J., Mattke, U., Strandberg, L., et al. (2001). The role of friction in the measurement of slipperiness, part 2: Survey of friction measurement devices. *Ergonomics, 44*(13), 1233–1261.

Chang, W. R., Kim, I. J., Manning, D. P., & Bunterngchit, Y. (2003a). *The role of surface roughness in the measurement of slipperiness. Measuring slipperiness—Human locomotion and surface factors* (pp. 101–117). Taylor & Francis.

Chang, W. R., Grönqvist, R., Leclercq, S., Brungraber, R. J., Mattke, U., Strandberg, L. et al. (2003b). *The role of friction in the measurement of slipperiness, Part 2: Survey of friction measurement devices. Measuring slipperiness—Human locomotion and surface factors* (pp. 135–163). Taylor & Francis.

Czichos, H. (1986). Introduction to friction and wear. In K. Friedrich (Ed.), *Friction and wear of polymer composites*. Amsterdam, The Netherlands: Elsevier.

Cohen, H. H., & Compton, D. M. (1982). Fall accident patterns: Characterization of most frequent work surface-related injuries. *Professional Safety, 27*(6), 16–22.

Courtney, T. K., Verma, S. K., Chang, W. R., Huang, Y. H., Lombardi, D. A., Brennan, M. J., et al. (2013). Perception of slipperiness and prospective risk of slipping at work. *Occupational and Environmental Medicine, 70*(1), 35–40.

Crowninshield, R. D., Johnston, R. C., Andrews, J. G., & Brand, R. A. (1978). A biomechanical investigation of the human hip. *Journal of Biomechanics, 11*(1–2), 75–85.

Currie, L. (2008). Fall and injury prevention, In R. G. Hughes (Ed.), *Patient safety and quality: An evidence-based handbook for nurses* (AHRQ Publication No. 08-0043). Agency for Healthcare Research & Quality.

Danion, F., Varraine, E., Bonnard, M., & Pailhous, J. (2003). Stride variability in human gait: The effect of stride frequency and stride length. *Gait & Posture, 18*(1), 69–77.

Diop, M., Rahmani, A., Belli, A., Gautheron, V., Geyssant, A., & Cottalorda, J. (2004). Influence of speed variation and age on the asymmetry of ground reaction forces and stride parameters of normal gait in children. *Journal of Pediatric Orthopaedics B, 13*(5), 308–314.

Donald, I. P., Pitt, K., Armstrong, E., & Shuttleworth, H. (2000). Preventing falls on an elderly care rehabilitation ward. *Clinical Rehabilitation, 14*(2), 178–185.

Dutruel, F., & Degas, G. (1997). Experimental analysis of main test devices used for evaluation of skid resistance. In P. Seppala, T. Luopaja, C. H. Nygård, & M. Mattila (Eds.), *Experience to innovation* (Vol. 3, pp. 371–373)., Helsinki Finnish Institute of Occupational Health: Finland.

Dykes, P. C., Carroll, D. L., Hurley, A. C., Benoit, A., & Middleton, B. (2009). Why do patients in acute care hospitals fall? Can falls be prevented? *The Journal of the American Medical Association, 39*(6), 299–304.

Ekkubus, C. F., & Killey, W. (1973). Validity of 0.5 static coefficient of friction (James Machine) as a measure of safe walking surfaces. *Soap/Cosmetics/Chemical Specialties, 49*(2), 40–45.

English, W. (1990). Improved tribometry on walking surfaces. In B. E. Gray (Ed.), *Slips, stumbles, and falls: Pedestrian footwear and surfaces, ASTM STP 1103* (pp. 73–81). Philadelphia: American Society for Testing and Materials.

Evans, D., Hodgkinson, B., Lambert, L., & Wood, J. (2001). Falls risk factors in the hospital setting: A systematic review. *International Journal of Nursing Practice, 7*(1), 38–45.

Expert Panel. (2004). *Reducing the number of slips/falls in school district facilities.* British Columbia School Safety Association (BCSSA), British Columbia Public School Employer's Association (BCPSEA) and the Workers' Compensation Board of British Columbia (WCB of BC), August. Available from: http://www.bcpsea.bc.ca/documents/expertpanel3.pdf.

Fendley, A. E., & Marpet, M. I. (1996). Required coefficient of friction versus walking speed: Potential influences of footwear and walkway surfaces. *Journal of Testing and Evaluation, 24*(6), 359–367.

Frenkel-Toledo, S., Giladi, N., Peretz, C., Herman, T., Gruendlinger, L., & Hausdorff, J. M. (2005). Effect of gait speed on gait rhythmicity in Parkinson's disease: Variability of stride time and swing time respond differently. *Journal of Neuro Engineering and Rehabilitation, 2*(23), 1–7.

Gadomski, N. (1998). Tread safely: Get a grip on slips, trips, and falls. *Safety & Health, 158*(3), 62(5).

Gao, C., & Abeysekera, J. (2004). A systems perspective of slip and fall accidents on icy and snowy surfaces. *Ergonomics, 47,* 573–598.

Giles, C., Saby, B., & Cardew, K. (1964). *Development and performance of the portable skid-resistance tester* (26 pp.). Department of Scientific and Industrial Research Road Research Laboratory Technical Paper No. 66, Her Majesty's Stationery Office, United Kingdom.

Greater London Council (GLC). (1971). *Testing and instruments—Pendulum.* GLC Bulletin No. 43, March. London, United Kingdom.

Greater London Council (GLC). (1985). Testing and instruments—Pendulum. GLC Bulletin 145, February. London, United Kingdom.

Grönqvist, R., Roine, J., Järvinen, E., & Korhonen, E. (1989). An apparatus and a method for determining the slip resistance of shoes and floors by simulation of human foot motions. *Ergonomics, 32*(8), 979–995.

Grönqvist, R., Hirvonen, M., & Tuusa, A. (1993). Slipperiness of the shoe-floor interface: Comparison of objective and subjective assessments. *Applied Ergonomics, 24*(4), 258–262.

Grönqvist, R., Hirvonen, M., & Tohv, A. (1998). Control of floor slipperiness: Evaluation of three portable test devices. In S. Kumar (Ed.), *Advances in occupational ergonomics and safety* (Vol. 2, pp. 96–99). Amsterdam, Netherlands: IOS Press.

Grönqvist, R., Hirvonen, M., & Tohv, A. (1999). Evaluation of three portable floor slipperiness testers. *Industrial Journal of Industrial Ergonomics, 25*(1), 85–95.

Grönqvist, R., Chang, W. R., Courtney, T. K., Leamon, T. B., Redfern, M. S., & Strandberg, L. (2001). Measurement of slipperiness: Fundamental concepts and definitions. *Ergonomics,* 44(13), 1102–1117.

Gray-Miceli, D., Ratcliffe, S. J., Liu, S., Wantland, D., & Johnson, J. (2012). Orthostatic hypotension in older nursing home residents who fall-are they Dizzy? *Clinical Nursing Research, 21*(1), 64–78.

Hamill, J., Van Emmerik, R. E. A., Heiderscheit, B. C., & Li, L. (1999). A dynamical systems approach to lower extremity running injuries. *Clinical Biomechanics, 14*(5), 297–308.

Hanson, J. P., Redfern, M. S., & Mazumdar, M. (1999). Predicting slips and falls considering required and available friction. *Ergonomics, 42*(12), 1619–1633.

Harris, G. W., & Shaw, S. R. (1988). Slip resistance of floors: User's opinions, Tortus instrument readings and roughness measurement. *Journal of Occupational Accidents, 9*(4), 287–298.

Hartikainen, S., Lonnroos, E., & Louhivuori, K. (2007). Medication as a risk factor for falls: Critical systematic review. *Journals of Gerontology. Series A: Biological Sciences and Medical Sciences, 62A*(10), 1172–1181.

Hathaway, J., Walsh, J., Lacey, C., & Saenger, H. (2001). Insights obtained from an evaluation of a falls prevention program set in a Rural Hospital. *Australian Journal of Rural Health, 9*(4), 172–177.

Hatton, A. L., Menant, J. C., Lord, S. R., Lo, J. C., & Sturnieks, D. L. (2013). The effect of lower limb muscle fatigue on obstacle negotiation during walking in older adults. *Gait & Posture, 37*(4), 506–510.

Hausdorff, J. M. (2004). Stride variability: Beyond length and frequency. *Gait & Posture, 20*(3), 304.

Hausdorff, J. M., Cudkowicz, M. E., & Firtion, R. (1998). Gait variability and basal ganglia disorders: Stride-to-stride variations of gait cycle timing in PD and Huntington's. *Movement Disorders, 13*(3), 428–437.

Health and Safety Executive (HSE). (2009). *The 3 ramp tests outlined in British and European standards*. England, UK: HSE.

Helbostad, J. L., Sturnieks, D. L., Menant, J., Delbaere, K., Lord, S. R., & Pijnappels, M. (2010). Consequences of lower extremity and trunk muscle fatigue on balance and functional tasks in older people: A systematic literature review. *BioMed Central Geriatrics, 10*(56), 1–8.

Herman, R., Wirta, R., Bampton, S., & Finley, F. R. (1976). Human solutions for locomotion: Single limb analysis. In R. M. Herman, S. Grillner, P. S. G. Stein, & D. G. Stuart (Eds.), *Neural control of locomotion*. New York: Plenum Press.

Hirvonen, M., Leskinen, T., Grönqvist, R., & Saario, J. (1994). Detection of near accidents by measurement of horizontal acceleration of the trunk. *Industrial Journal of Industrial Ergonomics, 14*(4), 307–314.

Hitcho, E. B., Krauss, M. J., Birge, S., Dunagan, W. C., Fischer, I., Johnson, S., et al. (2004). Characteristics and circumstances of falls in a hospital setting: A prospective analysis. *Journal of General Internal Medicine, 19*(7), 732–739.

Hoang, K., Stevenson, M. G., & Wilgoss, R. A. (1985). Measurement of dynamic friction between shoe soles and walkway surfaces. *Proceedings of the 22nd Annual Conference of the Ergonomics Society of ANZ* (pp. 265–271). Toowoomba.

Hoang, K., Stevenson, M. G., Nhieu, J., & Bunterngchit, Y. (1987). *Dynamic friction at heel strike between a range of protective footwear and non-slip floor surfaces*. Centre for Safety Science, University of New South Wales, Report No. CSS/1/87, July.

Hoehn, M. M., & Yahr, M. D. (1998). Parkinsonism: Onset, progression, and mortality. *Neurology, 50*(2), 318.

Huang, S. C., Chen, T. J., & Chou, P. (2005). Cumulative incident rate and associated factors of falls among the elderly in Shih-Pai, Taiwan. *Taiwan Journal of Public Health, 24*, 136–145.

International Standards Organization (ISO). (1992). *Footwear for professional use—Measurement of slip resistance*. Document ISO/TC 94/SC3, N162, November.

Jenson, A., & Chenoweth, H. H. (1990). *Applied engineering mechanics* (p. 164). Ohio: Glencoe/McGraw-Hill.

Jones, C., Manning, D. P., & Bruce, M. (1995). Detecting and eliminating slippery footwear. *Ergonomics, 38*(2), 242–249.

Jung, K., & Schenk, H. (1988). Objectification and accuracy of the walking method for determining the anti-slip properties of floor surfaces, (in German) Zentralblatt for Industrial Medicine, Germany. *Accident Prevention and Ergonomics, 39*(8), 221–228.

Kim, I. J. (1996a). Tribological concepts for the investigation of the pedestrian slipping and falling accidents—Part I. *International Occupational Injury Symposium* (p. 101), February, Sydney, Australia.

Kim, I. J. (1996b). Tribological approach for the analysis of pedestrian slip hazard—II. In *Proceedings of the '96 Spring Conference of Korean Institute of Industrial Engineers* (pp. 279–285). April, Seoul, Korea.

Kim, I. J. (1996c). Microscopic investigation to analyze the slip resistance of shoes. *Proceedings of the 4th Pan Pacific Conference on Occupational Ergonomics* (pp. 68–73). November, Taiwan, ROC.

Kim, I. J. (1996d). Microscopic observation of shoe heels for pedestrian slip hazard investigation. In *Proceedings of the 1st Annual International Conference on Industrial Application and Practice* (pp. 243–250). December, Texas, U.S.A.

Kim, I. J. (2002a). Identification of wear factors in the slip resistance measurements. *XVI International Annual Occupational Ergonomics and Safety Conference*. June, Toronto, Canada, CD-Rom.

Kim, I. J. (2002b). A new strategic study for the analysis of pedestrian slipping accidents. *The World Congress of Korean and Korean Ethnic Scientists and Engineers*. July, Seoul, Korea, CD-Rom.

Kim, I. J. (2003a). Observation of the contact areas of the heel surface during dynamic slip resistance measures. *15th Triennial Congress of the International Ergonomics Association*, IEA 2003, 7th Ergonomic Society of Korea/Japan Ergonomic Society Joint Conference, Seoul, Korea, CD-Rom.

Kim, I. J. (2003b). A novel study on the correlation of the characteristics of contact area and average slope angle with dynamic friction coefficients. *15th Triennial Congress of the International Ergonomics Association*, IEA 2003, 7th Ergonomic Society of Korea/Japan Ergonomic Society Joint Conference. Seoul, Korea, CD-Rom.

Kim, I. J. (2004a). Development of a new analyzing model for quantifying pedestrian slip resistance characteristics: Part I—Basic concepts and theories. *Industrial Journal of Industrial Ergonomics, 33*(5), 395–401.

Kim, I. J. (2004b). Development of a new analyzing model for quantifying pedestrian slip resistance characteristics: Part II—Experiments and validations. *Industrial Journal of Industrial Ergonomics, 33*(5), 403–414.

Kim, I. J. (2006a). The current hiatus in fall safety measures. In W. Karwowski (Ed.), *International encyclopedia of ergonomics and human Factors-2005* (pp. 2572–2576). USA: Taylor & Francis Group, LLC.

Kim, I. J. (2006b). A new paradigm for characterizing slip resistance properties. In W. Karwowski (Ed.), *International encyclopedia of ergonomics and human factors-2005* (pp. 2735–2740). USA: Taylor & Francis Group, LLC.

Kim, I. J. (2015a). Practical design search for optimal floor surface finishes to prevent fall incidents. In B. Evans (Ed.), *Accidental falls: Risk factors, prevention strategies and long-term outcomes* (Chapter 5, pp. 80–103). Hauppauge, NY 11788, USA: Nova Science Publishers, Inc.

Kim, I. J. (2015b). Slip-resistance measurements for assessing pedestrian falls: Facts and fallacies. In B. Evans (Ed.), *Accidental falls: Risk factors, prevention strategies and long-term outcomes* (Chapter 6, pp. 105–125). Hauppauge, NY 11788, USA: Nova Science Publishers, Inc.

Kim, I. J. (2015c). Wear observation of shoe surfaces: Application for slip and fall safety assessments. *Tribology Transactions, 58*(3), 407–417.

Kim, I. J. (2015d). Research challenges on slip-resistance measurements for assessing pedestrian fall incidents. *Journal of Ergonomics, 5*(3). doi:10.4172/2165-7556.1000e142

Kim, I. J. (2016a). A study on wear development of floor surfaces: Impact on pedestrian walkway slip-resistance performance. *Tribology International, 95*, 316–323.

Kim, I. J. (2016b). Identifying shoe wear mechanisms and associated tribological characteristics: The importance for slip resistance evaluation. *Wear, 360–361,* 77–86.

Kim, I. J., Smith, R. (1998a). A study of the comparative geometry mating between the surfaces of the shoe and floor in pedestrian slip resistance measurements. *The 5th Pan-Pacific Conference on Occupational Ergonomics* (pp. 34–37), July, Kitakyushu, Japan.

Kim, I. J., & Smith, R. (1998b). Tribological characterization of the frictional force component in pedestrian slip resistance measurements. *Third World Congress of Biomechanics (WCB '98).* August, Hokkaido, Japan.

Kim, I. J., & Smith R. (1999). The relationship between wear, surface topography characteristics and coefficient of friction as a means of assessing the slip hazards. *2nd Asia-Pacific Conference on Industrial Engineering and Management Systems (APIEMS'99)* (pp. 155–161). October, Ashikaga, Japan.

Kim, I. J., & Smith, R. (2000). Observation of the floor surface topography changes in pedestrian slip resistance measurements. *Industrial Journal of Industrial Ergonomics, 26*(6), 581–601.

Kim, I. J., & Smith, R. (2001a). A critical analysis on the friction measuring concept for slip resistance evaluation. In A. S. T. M. Headquarters (Ed.), *ASTM Symposium on the Metrology of Pedestrian Locomotion and Slip Resistance, June* (pp. 1–14). USA: West Conshodocken, Pennsylvania.

Kim, I. J., & Smith, R. (2001b). A study for characterising topography changes of shoe surfaces in the early stage of slip resistance measurements—Bearing Area Curve. *6th Pan-Pacific Conference on Occupational Ergonomics* (pp. 299–303). August, Beijing, P. R. China.

Kim, I. J., & Smith, R. (2001c). Three-dimensional analysis of floor surface wear during slip resistance measurements. *6th Pan-Pacific Conference on Occupational Ergonomics* (pp. 304–308). August, Beijing, P. R. China.

Kim, I. J., & Smith, R. (2003). A critical analysis of the relationship between shoe-floor wear and pedestrian/walkway slip resistance. In M. I. Marpet & M. A. Sapienza (Eds.), *Metrology of Pedestrian Locomotion and Slip Resistance* (pp. 33–48). American Society of Testing and Materials, Special Technical Publication 1424, USA: ASTM International, Philadelphia.

Kim, I. J., & Nagata, H. (2008a). Nature of the shoe wear: Its uniqueness, complexity and effects on slip resistance properties. *Contemporary Ergonomics 2008, Taylor & Francis, 15,* 728–734.

Kim, I. J., & Nagata, H. (2008b). Research on slip resistance measurements—A new challenge. *Industry Health, 46*(1), 66–76.

Kim, I. J., & Alduhishy, A. (2013). Experimental evidences of floor wear and its applications for slip and fall safety evaluation. *Proceedings of the International Conference on Fall Prevention and Protection* (pp. 99–104). Tokyo, Japan.

Kim, I. J., Smith, R., & Nagata, H. (2001). Microscopic observations of the progressive wear on the shoe surfaces which affect the slip resistance characteristics. *Industrial Journal of Industrial Ergonomics, 28*(1), 17–29.

Kim, I. J., Hsiao, H., & Simeonov, P. (2013). Functional levels of floor surface roughness for the prevention of slips and falls: Clean-and-dry and soapsuds-covered wet surfaces. *Applied Ergonomics, 44*(1), 58–64.

Knoll, Z., Kocsis, L., & Kiss, R. M. (2004). Gait patterns before and after anterior cruciate ligament reconstruction. *Knee Surgery, Sports Traumatology, Arthroscopy, 12*(1), 7–14.

Krauss, M. J., Evanoff, B., Hitcho, E., Ngugi, K. E., Dunagan, W. C., Fischer, I., et al. (2005). A case-control study of patient, medication, and care-related risk factors for inpatient falls. *Journal of General Internal Medicine, 20*(2), 116–122.

Krauss, M. J., Nguyen, S. L., Dunagan, W. C., Birge, S., Costantinou, E., Johnson, S., et al. (2007). Circumstances of patient falls and injuries in 9 hospitals in a Midwestern Healthcare System. *Infection Control and Hospital Epidemiology, 28*(5), 544–550.

Kulakowski, B. T., Buczek, F. L., Cavanagh, P. R., & Pradhan, P. (1989). Evaluation of performance of three slip resistance testers. *Journal of Testing and Evaluation, 17*(4), 234–240.

Lanshammar, H., & Strandberg, L. (1983). Horizontal floor reaction forces and heel movements during the initial stance phase. In H. Matsui & K. Kobayashi (Eds.), *Biomechanics VIII* (pp. 1123–1128). Baltimore: University Park Press.

Layne, L. A., & Landen, D. D. (1997). A descriptive analysis of nonfatal occupational injuries to older workers, using a national probability sample of hospital emergency departments. *Journal of Occupational and Environmental Medicine, 39*(9), 855–865.
Leamon, T. B., & Son, D. H. (1989). The natural history of a micro slip. In A. Mital (Ed.), *Advances in industrial ergonomics and safety I* (pp. 633–638). Taylor & Francis.
Leamon, T. B., & Li, K. W. (1990). Microslip length and the perception of slipping. *Proceedings of 23rd International Congress on Occupational Health.* September 22–28, Montreal, Canada.
Leamon, T. B. (1992). The reduction of slip and fall injuries: Part II—The scientific basis (knowledge base) for the guide. *International Journal of Industrial Ergonomics, 10,* 29–34.
Leclercq, S., Tisserand, M., & Saulnier, H. (1995). Assessment of slipping resistance of footwear and floor surfaces. Influence of manufacture and utilization of the products. *Ergonomics, 38*(2), 209–219.
Leclercq, S. (1999). Prevention of the same level falls: A more global appreciation of this type of accident. *Journal of Safety Research, 30*(2), 103–112.
Leclercq, S., & Saulnier, H. (2002). Floor slip resistance changes in food sector workshops: Prevailing role played by "fouling". *Safety Science, 40,* 659–673.
Leung, A., Chi, I., Lou, V. W. Q., & Chan, K. S. (2010). Psychosocial risk factors associated with falls among Chinese community-dwelling older adults in Hong Kong. *Health and Social Care in the Community, 18*(3), 272–281.
Lockhart, T. E. (1997). *Biomechanics of Slips and Falls in the Elderly.* (PhD Thesis), Texas Tech University.
Lockhart, T. E., Woldstad, J. C., & Smith, J. L. (2003). Effects of age-related gait changes on the biomechanics of slips and falls. *Ergonomics, 12,* 1136–1160.
Lockhart, T. E., Smith, J. L., & Woldstad, J. C. (2005). Effects of ageing on the biomechanics of slips and falls. *Human Factors, 47*(4), 708–729.
Majcherczyk, R. (1978). A different approach to measuring pedestrian friction: The CEBTP skidmeter. In C. Anderson & J. Senne (Eds.), *Walkway surfaces: Measurement of slip RESISTANCE, ASTM STP 649* (pp. 88–99). Philadelphia: American Society for Testing and Materials.
Manning, D. P., Ayers, I., Jones, C., Bruce, M., & Cohen, K. (1988). The incidence of underfoot accidents during 1985 in a working population of 10,000 Merseyside people. *Journal of Occupational Accidents, 10*(2), 121–130.
Marigold, D. S., & Patla, A. E. (2002). Strategies for dynamic stability during locomotion on a slippery surface: Effects of prior experience and knowledge. *Journal of Neurophysiology, 88* (1), 339–353.
Marigold, D. S., Bethune, A. J., & Patla, A. E. (2003). Role of the unperturbed limb and arms in the reactive recovery response to an unexpected slip during locomotion. *Journal of Neurophysiology, 89*(4), 1727–1737.
Marosszeky, M. (1985). *Slip and abrasion resistance of clay brick pavers.* Building Research Centre, University of New South Wales, Report No. 3.85, November, Australia.
Marpet, M. I. (1996). On threshold values that separate pedestrian walkways that are slip resistant from those that are not. *Journal of Forensic Science, 41*(5), 747–755.
Maynard, W. (2006). Preventing outdoor same-level: Slips, trips, and falls: By properly maintaining sidewalks, parking lots, curbs and other outdoor pedestrian spaces, winter's slips, trips and falls can be reduced or eliminated. *Occupational Hazards, 68*(11), 41–43.
McElvaney, L. F. (1993). *Proceedings of the Sixth Southeastern Safety and Health Conference and Exhibition* (p. 227). Atlanta (GA): Georgia Tech Research Institute.
Moore, D. F. (1972). The friction and lubrication of elastomers. *International series of monographs on materials science and technology* (Vol. 9). Oxford: Pergamon Press Ltd., Headington Hill Hall.
Munro Instruments. (2013). *Portable skid resistance tester user manual.* Munro Instruments. p. 28.
Myung, R., Smith, J. L., & Leamon, T. B. (1992a). Slip distance as an objective criterion to determine the dominant parameter between static and dynamic COFs. *Proceedings of the Human Factors Society, 36th Annual Meeting* (pp. 738–741).

Myung, R., Smith, J. L., & Leamon, T. B. (1992b). Slip distance for slip/fall studies. *Advances in Industrial Ergonomics and Safety IV*, 983–987.

National Research Council. (1961). *Causes and measurement of walkway slipperiness*. Federal Construction Council, Technical Report No. 43. Washington, DC: National Academy of Sciences—NRC, Publication 899.

Nenonen, N. (2013). Analysing factors related to slipping, stumbling, and falling accidents at work: Application of data mining methods to finnish occupational accidents and diseases statistics database. *Applied Ergonomics, 44*(2), 215–224.

Oates, A. R., Patla, A. E., Frank, J. S., & Greig, M. A. (2004). Control of dynamic stability during gait termination on a slippery surface. *Journal of Neurophysiology, 93*(1), 64–70.

Owings, T. M., & Grabiner, M. D. (2004). Step width variability, but not step length variability or step time variability, discriminates gait of healthy young and older adults during treadmill locomotion. *Journal of Biomechanics, 37*(6), 935–938.

Pai, Y. C., Wening, J. D., Runtz, E. F., Iqbal, K., & Pavol, M. J. (2003). Role of feedforward control of movement stability in reducing slip-related balance loss and falls among older adults. *Journal of Neurophysiology, 90*(2), 755–762.

Park, K. (2012). *Quantitative assessment of human gait patterns using biomechanical and dynamical tools* (PhD Thesis). University of Illinois at Urbana-Champaign, USA.

Pavol, M. J., Runtz, E. F., & Pai, C. Y. (2004). Diminished stepping responses lead to a fall following a novel slip induced during a sit-to-stand. *Gait & Posture, 20*(2), 154–162.

Pearson K. B., & Coburn, A. F. (2011). *Evidence-based falls prevention in critical access hospitals*. Policy Brief #24. December. Available at: http://www.flexmonitoring.org/wp-content/uploads/2013/07/PolicyBrief24_Falls-Prevention.pdf

Peeters, G., van Schoor, N. M., & Lips, P. (2009). Fall risk: The clinical relevance of falls and how to integrate fall risk with fracture risk. *Best Practice & Research Clinical Rheumatology, 23*(6), 797–804.

Perkins, P. J. (1978). Measurement of slip between the shoe and ground during walking. In C. Anderson & J. Seene (Eds.), *Walkway surfaces: Measurement of slip resistance, ASTM STP 649* (pp. 71–87).

Perkins, P. J., & Wilson, M. P. (1983). Slip resistance testing of shoes—New developments. *Ergonomics, 26*(1), 73–82.

Perry, J., & Burnfield, J. (2010). *Gait analysis: Normal and pathological function*. Slack Incorporated; Second Edition: Thorofare, NJ.

Powers, C. M., Brault, J. R., Stefanou, M. A., Tsai, Y. J., Flynn, J., & Siegmund, G. P. (2007). Assessment of walkway tribometer readings in evaluating slip resistance: A gait-based approach. *Journal of Forensic Sciences, 52*(2), 400–405.

Redfern, M. S., & Bidanda, B. (1994). Slip resistance of the shoe-floor interface under biomechanically relevant conditions. *Ergonomics, 37*(3), 511–524.

Redfern, M. S., Cham, R., Gielo-Perczak, K., Grönqvist, R., Hirvonen, M., Lanshammar, H., et al. (2001). Biomechanics of slips. *Ergonomics, 44*(13), 1138–1166.

Reed, M. E. (1975). *Standardization of friction testing of industrial working surfaces* (U.S. Department of Health, Education, and Welfare, Publication No. NIOSH 76–123). Cincinnati Ohio: Government Printing Office.

Regan Scientific Instruments. (2015). *BOT-3000E Digital Tribometer Operating Manual*. Regan Scientific Instruments.

Rhalimi, M., Helou, R., & Jaecker, P. (2009). Medication use and increased risk of falls in hospitalized elderly patients: A retrospective, case-control study. *Drugs & Aging, 26*(10), 847–852.

Rhoades, T. P., & Miller, J. M. (1988). Measurement and comparison of 'Required' versus 'Available'. *Proceedings of the 21st Annual Conference of the Human Factors Association of Canada* (pp. 137–140). Edmonton, Alberta, Canada.

Ricotti, R., Delucchi, M., & Cerisola, G. (2009). A comparison of results from portable and laboratory floor slipperiness testers. *International Journal of Industrial Ergonomics, 39*, 353–357.

Rolfsman, E., Bylund, P. O., & Saveman, B. I. (2012). Single injury incidents among pedestrians and bicyclists in northern Sweden—Safety and preventive issues. *Safety Science Monitor, 16*(1), 1–10.

Rubenstein, L. Z., & Josephson, K. R. (2006). Falls and their prevention in elderly people: What does the evidence show? *Medical Clinics of North America, 90*(5), 807–824.

von Schroeder, H. P., Coutts, R. D., Lyden, P. D., Billings, E., Jr., & Nickel, V. L. (1995). Gait parameters following stroke: A practical assessment. *Journal of Rehabilitation Research and Development, 32*(1), 25–31.

SAI Global. (2013). *AS4586-2013—Slip resistance classification of new pedestrian surface materials*. Available at: http://infostore.saiglobal.com/store/details.aspx?ProductID=1636572

Schieb, D. A., Lochocki, R. F., & Gautschi, G. H. (1990). Frictional and ground reaction force measurement with the Multicomponent Quartz Force Plate. In B. E. Gray (Ed.), *Slips, stumbles, and falls: Pedestrian Footwear and surface, ASTM STP 11031* (pp. 102–112). Philadelphia: American Society for Testing and Materials.

Schwartz, M. H., Rozumalski, A., & Trost, J. P. (2008). The effect of walking speed on the gait of typically developing children. *Journal of Biomechanics, 41*(8), 1639–1650.

Sigler, A., Geib, M. N., & Boone, T. H. (1948). "Measurement of the slipperiness of walkway surfaces", Research paper RP 1879. *Journal of Research of the National Bureau of Standards, 40,* 339–346.

Skiba, R., Wortmann, H. R., & Mellwig, D. (1985). Bewegungsabläufe und Kräfte beim Treppenaufstieg un-abstieg aus der Sicht der Gleitsicherheit. *Zeitschrift für Arbeitswissenschaft* 11, 97–100 (in German with English summary).

Skoyles, J. R. (2006). Human balance, the evolution of bipedalism and dysequilibrium síndrome. *Medical Hypotheses, 66*(6), 1060–1068.

SlipAlert. (2011). Available at: http://www.slipalert.com/Home/about-slipalert.htm

Sotter, G. (2000). *STOP slip and fall accidents!* CA, USA: Sotter Engineering Corporation.

Sotter Engineering Corporation. (2017). Slip resistance testing standards in 2017. Mission Viejo, CA USA: Sotter Engineering Corporation. Available at: https://safetydirectamerica.com/slipresistancetestingstandards

Standards Australia, "HB 197:1999. *Standards Australia 1999.* Australia.

Standards Centre. (2010). *Screeds, bases and in situ floorings. Synthetic resin floorings. Code of practice.* BS 8204-6:2008+A1:2010. UK: Standards Centre.

Staal, C., White, B., Brasser, B., LeForge, L., Dlouhy, A., & Gabier, J. (2004). Reducing employee slips, trips, and falls during employee-assisted patient activities. *Rehabilitation Nursing: The Official Journal of the Association of Rehabilitation Nurses, 29*(6), 211–214.

Stevenson, M. G., Hoang, K., Bunterngchit, Y., & Lloyd, D. G. (1989). Measurement of slip resistance of shoes on floor surfaces, Part 1: Methods. *Journal of Occupational Health and Safety—Aust NZ, 5*(2), 115–120.

Stevenson, M. G. (1997). Evaluation of the slip resistance of six types of women's safety shoe using a newly developed testing machine. *Journal of Occupational Health and Safety—Aust NZ, 13*(2), 175–182.

Strandberg, L. (1983). On accident analysis and slip resistance measurement. *Ergonomics, 26*(1), 11–32.

Strandberg, L., & Lanshammar, H. (1981). The dynamics of slipping accidents. *Journal of Occupational Accidents, 3*(3), 153–162.

Strandberg, L., & Lanshammar, H. (1983). On the biomechanics of slipping accidents. In H. Matsui & K. Kobayashi (Eds.), *Biomechanics VIII* (pp. 397–402). Baltimore: University Park Press.

Suh, N. P., & Sin, H. C. (1981). The genesis of friction. *Wear, 69*(1), 91–114.

The Joint Commission. (2007). Improving fall risk assessment. *Good practices in preventing patient falls: A collection of case studies* (pp. 17–29). Oakbrook Terrace, IL: Joint Commission Resources.

Tiberio, D. (1987). The effect of excessive subtalar joint pronation on patellofemoral mechanics: a theoretical model, *Journal of Orthopaedic and Sports Physical Therapy, 9*(4), 160–165.

Tinetti, M. E., & Kumar, C. (2010). The patient who falls. *Journal of the American Medical Association, 303*(3), 258–266.

Tisserand, M. (1985). Progress in the prevention of falls caused by slipping. *Ergonomics, 28*(7), 1027–1042.

Titler, M. G., Shever, L. L., Kanak, M. F., Picone, D. M., & Qin, R. (2011). Factors associated with falls during hospitalization in an older adult population. *Research and Theory for Nursing Practice, 25*(2), 127–148.

Tzeng, H. M. (2010a). Inpatient falls in adult acute care settings: Influence of patients' mental status. *Journal of Advanced Nursing, 66*(8), 1741–1746.

Tzeng, H. M. (2010b). Understanding the prevalence of inpatient falls associated with toileting in adult acute care settings. *Journal of Nursing Care Quality, 25*(1), 22–30.

UK Slip Resistance Group (UKSRG). (2016). *The assessment of floor slip resistance*. Coventry, UK: The UK Slip Resistance Group Guidelines.

Verma, S. K., Chang, W. R., Courtney, T. K., Lombardi, D. A., Huang, Y. H., Brennan, M. J., et al. (2011). A prospective study of floor surface, shoes, floor cleaning and slipping in US limited-service restaurant workers. *Occupational and Environmental Medicine, 68*(4), 279–285.

Wade, D. T., Wood, V. A., Heller, A., Maggs, J., & Langton, H. R. (1987). Walking after stroke: Measurement and recovery over the first 3 months. *Scandinavian Journal of Rehabilitation Medicine, 19*(1), 25–30.

Wilson, M. P. (1990). Development of SATRA slip test and tread pattern design guideline. In B. E. Gray (Ed.), *Slips, stumbles, and falls: Pedestrian footwear and surfaces, ASTM STP 1103* (pp. 113–123). Philadelphia: American Society for Testing and Materials.

Whittle, M. W. (1999). Generation and attenuation of transient impulsive forces beneath the foot: A review. *Gait & Posture, 10*(3), 264–275.

Weisberger, K. (1994). Taking action against loss of traction means maintenance, training, footwear. *Occupational Health & Safety, 63*(10), 139.

You, J., Chou, Y., Lin, C., & Su, F. (2001). Effect of slip on movement of body centre of mass relative to base of support. *Clinical Biomechanics, 16*(2), 167–173.

Yu, P. L., Qin, Z. H., Shi, J., Zhang, J., Xin, M. Z., Wu, Z. L., et al. (2009). Prevalence and related factors of falls among the elderly in an urban community of Beijing. *Biomedical and Environmental Sciences, 22*(3), 179–187.

Zijlstra, G. A., van Haastregt, J. C., van Eijk, J. T., van Rossum, E., Stalenhoef, P. A., & Kempen, G. I. (2007). Prevalence and correlates of fear of falling, and associated avoidance of activity in the general population of community-living older people. *Age and Ageing, 36*(3), 304–309.

Chapter 3
Pedestrian Slip Resistance Measurements: Verities and Challenges

3.1 Introduction

Fall incidents resulting from slips or trips are one of the most common injuries at workplaces. On average, they cause 40% of all reported major injuries and lead to other types of serious accidents such as falls from height (HSE 2011). Slips and falls are also one of the leading categories of non-traffic accidents in terms of serious injuries and fatalities, as well as one of the leading causes of injury-related deaths for the elderly aged 65 and over (Layne and Landen 1997; Scott 2009; Yeoh et al. 2013).

There have thus been prolonged efforts to understand the main causes of such incidents throughout the world. Abundance researches in the literature have shown that the most common precipitating event leading to a fall is a loss of traction or slip resistance between the shoe sole/heel and floor surface. As a result, its slip resistance or traction property has been commonly measured as a form of a coefficient of friction (COF). Hence, knowledge about friction demand and the friction available has been recognised as one of the key factors for the fall safety assessment.

Despite many years of investigations with fabrication of numerous testing apparatuses and devices, none of them is internationally adopted as a standard model and a device for slip resistance measurements, because each of them has some advantages and disadvantages in its design and testing performance (Kim 2006, 2015a; Kim and Nagata 2008a, b). In fact, each testing instrument has different concepts, measuring parameters and mechanical principles so it would be unreasonable to adopt a reference value without any citation to the instrument and/or device used for slip resistance measurements. This means that there is a great deal of uncertainties about what constitutes a "safe" quantity for the COF between the footwear and underfoot surfaces.

I.-J. Kim, *Pedestrian Fall Safety Assessments*,
DOI 10.1007/978-3-319-56242-1_3

Although the concept of friction is relatively simple and straightforward, solving the real-world problems on slip and fall incidents are a quite complex and challenging task (Kim et al. 2013; Kim 2015b, d). However, one of the most crucial aspects to address is that the COF is not a constant quantity. Because initial surface features and topographic characteristics of both shoes and floors are frequently and significantly modified from the first moment of contact by repetitive friction and wear developments. As a result, frictional properties become noisy and continuously change as a function of a complex array of tribo-physical phenomena amongst the shoes, floors, and environments (Kim and Smith 2000, 2003; Kim 2006, 2015a, b; Kim and Nagata 2008a, b).

Surprisingly, however, there are almost no studies on how friction induced wear developments of the shoe heel and floor surface to affect slip resistance performance. In addition, slip resistance properties observed at the sliding interface between the shoe and floor surfaces are diverse and combine various sub-mechanisms of friction and wear events (Kim and Smith 2000, 2003; Kim 2004a, 2006, 2015a, b; Kim and Nagata 2008a, b). Thus, a simple format of friction measurement does not provide an accurate determination of essential slip-resistance properties between the shoe and floor and accordingly has obvious difficulty as an indicator for the fall safety assessments.

In this sense, a fresh insight is required to systematically handle multi-factorial characteristics of friction and wear behaviours of the shoe and floor and their interactions on slip resistance performance. This practice would provide more reliable measurement results on slip resistance properties than a commonly practised simple COF reading between the shoe and floor. Whilst controversies around the friction measurement for slipperiness assessment still remain (Chang et al. 2001; Kim 2004a, b, 2015a; Kim and Nagata 2008a, b), a tribological approach may provide an objective alternative way to overcome the current limitations on slip resistance evaluations.

Therefore, this chapter robustly discusses the limitations on the commonly practised present concept on slip resistance measurements as a simple format of COF quantity. This chapter also explores the seriousness of misinterpretations on slip resistance measurement results that are mainly caused by oversimplified views on frictional behaviours between the shoe heels and floor surfaces. Based on such critical reviews and discussions, a new paradigm for the analysis of slip resistance properties between the shoe and floor is proposed for the future researches on the fall safety evaluations.

3.2 Brief Overview

One of the most important aspects is that a COF quantity is not a constant value and continuously changes during slip resistance measurements (Kim and Smith 1998a, b, 2013; Kim and Smith 2000; Kim 2015c). This means that frictional behaviours observed at a sliding interface between any footwear and floor surface show diverse and combine various sub-mechanisms of friction and wear events (Kim and Smith 2000, 2003; Kim 2004a).

Moreover, from a geometrical point of view, almost all surfaces are rough on a microscopic scale and are comprised of an aggregation of micro- and macro-asperities (Kim 2015b). These topographic characteristics of both shoe and floor surfaces are continually modified during sliding processes. That is, initial surface features and properties of both bodies are significantly changed and continuously modified from the first moment of contact by surface failures and wear evolutions (Proctor and Coleman 1988; Kim and Smith 2000, 2003; Kim et al. 2001, 2013; Kim 2004a, 2015c, 2016; Kim and Nagata 2008a, b). Therefore, a simple format of friction measurement may not be good enough to provide an accurate measure of slip resistance properties between the footwear and underfoot surfaces. In spite of the importance of this matter, this issue has not been studied in depth.

In order to analyse how the topographies of a randomly rough floor surface interact with the ones of a shoe heel under a load, it is necessary to study how the frictional force is generated. Since slip resistance properties are significantly related to the surface features, it can be expected that the friction mechanism between the shoe heel and floor surface is accordingly affected by surface alterations and wear evolutions during the repeated sliding friction process (Kim et al. 2013; Kim 2015b). Whilst the COFs vary in a complex and non-uniform way during continuous ambulation, it is found that initial heel contact conditions are progressively changed with wear developments and this result directly influences on the changes of the frictional force component (Kim 2015c, 2016).

This fact indicates that the COF is not a constant for any pair of the shoe-floor combination, sliding against each other under a given set of surface and environmental conditions, especially in the dry surface condition. For the improved analysis of multifaceted aspects of slip resistance properties between the footwear and underfoot surfaces, a further in-depth knowledge is necessary to understand complex mechanisms of tribolo-physical characteristics.

Therefore, one of the primary objectives of this chapter is to study this issue and critically analyse tribological features of the frictional force generated between the shoe heels and floor surfaces during sliding friction measurements.

For example, the following issues should be systematically measured, monitored, and analysed:

(1) the contact time and rate of increase of the frictional force,
(2) the heel contact angles at the time of maximum friction demand, and
(3) the surface roughness of both shoe and floor surfaces.

To perceive the fundamental aspects of friction and wear behaviours and comprehend related tribo-physical characteristics between the shoe and floor, it would be beneficial to control test conditions of slip resistance measurements such as testing under clean and dry environments as an initial step.

This approach is intended to eliminate any confounding effect of surface contaminants by other than the shoe and floor itself. This practice may provide clear ideas how topographic alterations of both shoe and floor surfaces are progressed during sliding friction events.

Because we are more often exposed to the clean and dry surface conditions rather than lubricated and contaminated ones, it also can be considered that studies on slip resistance properties under the clean and dry environments seem to be a worthwhile attempt to identify principal mechanisms of tribo-physical characteristics of the two bodies and their effects on slip resistance performance between them.

Acknowledging that lubricated floorings are more potential slip and fall hazards than dry ones, the traction behaviours and mechanisms between the two surface conditions are drastically different. Without thorough understanding of their tribological properties, therefore, any simplified friction measurement is meaningless as long as we know the main problems on slip resistance measurements.

3.3 Theoretical Backgrounds

As demonstrated above, friction and wear events are the results of extremely complex interactions between the surface and near-surface regions of the two sliding materials. These regions differ from the bulk material characteristics in their physical, chemical, and mechanical properties. Furthermore, these properties can change radically because of interactions of the surface atoms with their environments and with each other.

If it is necessary to measure a COF between a pair of materials under a specific set of operating conditions, then the safest procedure is to simulate experimentally, under conditions as close to the operating conditions as is feasible. Hence, the following important issues should be considered to design slip resistance measurements:

(1) a COF is not a constant for any pairing of materials that may be similar or dissimilar, sliding against each other under a given set of surfaces and environmental conditions;
(2) surface topographies are complex and the characteristics of both mating surfaces are continuously changing in the process of sliding friction induced wear and tear growths; and
(3) selection of test conditions for friction measurement influences the slip resistance performance.

Strandberg and Lanshammar suggested that the vertical force was one of the particularly important variables that slip resistance meters should reproduce based on tribological and practical experience (Strandberg and Lanshammar (1983). Of many important parameters related to crucial gait phase, Redfern and Bidanda classified the vertical force, shoe heel velocity, and shoe contact angle as the most significant biomechanical parameters (Redfern and Bidanda 1994).

Grönqvist et al. (1989) and Wilson (1990) recommended that the range of vertical force in the slip testers varied from almost a whole-body weight to 50% of a body weight. However, in a recent study, Chaffin et al. (1992) argued that although the basic assumption in computing the COF was that the friction force increased proportionally to the normal force holding the contacting surfaces together under some circumstances, this might not be true.

However, the effect of the normal load is not simple to predict. For example, when a soft shoe material such as a crepe is loaded heavily on a rough surface, the effect of deformation force on friction seems to be larger than when using a smooth surface or a harder shoe material (Chaffin et al. 1992). That is, an interaction between the shoe heel/sole and floor surface exists so this complicates to generalise the COFs between under a light and a heavy load. This fact is particularly troublesome when selecting a slip tester because some COF testing procedures use a very light normal load (less than 10 N), but others advocate a quite heavy load (about 1000 N) for the normal one.

The angle at which the shoe heel first contacts the floor surface seems to be an important factor in a slipping mechanism. Hence, striking at a "correct angle" would mean testing one of the most important parts of the shoe in walking. A motor driven treadmill was built to investigate the human walking patterns and more importantly the variations of strike angles with different walking speeds and subjects (Hoang et al. 1987). The treadmill was designed to simulate walking on a flat surface, and up or down a ramp. Its speed was continuously controlled within the range of 0.3–2.2 m/s.

An individual video system was used to record the walking patterns and the strike angles were measured off the TV screen in a playback pause mode. Hoang et al. (1987) found that the heel strike angle varied with the walking speed, step size, and subject's height. On a horizontal surface, it laid within the range of 6–10° measured from the floor. Analysing the distribution, a mean strike angle of 9° was chosen for their dynamic friction tests.

Given that vertical and normal loads and strike angle are concerned, it remains to select the test speed. According to Hoang et al. (1987), this would not be an important issue where the COF was found not to vary much with the test speed. However, this decision seems not to be the case, as shown by Perkins and Wilson (1983). The need to include the test speed as a variable would add tremendously to the volume of tests to assess the merits of shoe and floor materials.

Also, users of the data would then need to consider which test speed is appropriate. It would be, therefore, desirable to identify a test speed which is representative of slipping risk and keep this constant during tests. At present, however, the issue of the testing speed is still the subject of debate amongst fall safety researchers. Walking speed varies from 1 to 2 m/s. but, a heel edge forward speed is likely to be considerably less than this just before the heel strike.

Strandberg (1983) reported experimental results for different subjects, in which test speeds varied from 0.06 to 1.7 m/s. He also found the heel velocity at skid start varied from 0.08 to 0.32 m/s. After skid started, the shoe heel accelerated to a value above that of walking speed. He concluded that the speed of a dynamic test should be in the range of 0–0.5 m/s. Perkins and Wilson (1983) stated from similar measurements that "probably the ideal speed for high-speed measurement is 0.5–1.0 m/s. since the foot and shoe can be travelling at this speed when the heel tip contacts the ground. Even if slip starts from a static situation, such is the acceleration that the speed of slip is about 0.5 m/s. after only 0.01 m slip distance". The issues of slip distance were also studied by Myung et al. (1992a, b).

On the other hand, commercially available slip testers have adopted quite different test speeds in their devices. For example, in the case of the Tortus floor friction tester, it uses a speed of 0.017 m/s. This test speed is very low in comparison with the above recommendations, but the developers show that the effect of speed is not great for the floors tested (Harrison and Malkin 1983). Whilst the Sigler swinging pendulum tester uses a test speed of about 2.7 m/s (James 1983).

Dynamic friction testers used in this study were designed for an adjustable speed up to 0.6 m/s. In a preliminary series of tests, it was found that there was little variation of the results with speed (less than 8%) within the speed range of 0.2–0.6 m/s (Hoang et al. 1985). Therefore, the test speed of 0.4–0.5 m/s was chosen in this book.

3.4 Mislead Issues on Slip Resistance Measurements

A great deal of research has been undertaken for the prevention of pedestrian fall incidence. Many different theories, devices and experimental set-ups have been developed to establish safety criteria in terms of static friction coefficient (SFC) as well as dynamic friction coefficient (DFC), or slip resistance index amongst a diverse range of shoes, floor, and environments (Kim and Smith 2003). Thus, most studies from a variety of slip resistance testers and apparatuses have referenced their own COF quantities as a safety threshold for the assessment of fall risks and fall incidents. However, this approach seems to have serious limitations. Because each testing device or apparatus has different theoretical concepts and technical principles, it would be unreasonable to adopt the reference COF value without any citation to the instrument employed for the slip resistance measurements. This means that there is a great uncertainty about what a "safe" index or value for the COF between the footwear and underfoot surface should be.

Another critical aspect to highlight is that the COF index or quantity is not a constant value and continuously changes during slip resistance measurements (Kim and Smith 1998a, b, 2013; Kim and Smith 2000; Kim 2015a). Topographic features of both shoe andfloor surfaces are constantly altered by the course of combined effect of friction and wear developments. This means that initial surface finishes and topographic structures are significantly modified and continuously changed from the first moment of contact by surface failures and wear progresses (Proctor and Coleman 1988; Kim and Smith 2000, 2003; Kim et al. 2001, 2013; Kim 2004a, 2015a, b; Kim and Nagata 2008a, b).

In addition, friction phenomena observed at a sliding interface between any footwear and floor surface are diverse and combine various sub- mechanisms (Leclercq et al. 1993; Grönqvist 1995; Kim and Smith 2000, 2003; Kim et al. 2013; Kim 2015a). Therefore, a simple format of friction or traction measurement may not be accurate enough to provide true measures of slip resistance properties between the footwear and underfoot surface. Despite the importance of those complex issues, this matter has not been systematically analysed in the literature. Hence, it would be beneficial to review the definition of friction and redefine it from a fresh point of view.

3.5 Definition of a COF

As a well-known quantity, a coefficient of friction (COF) has long been used as a fall safety indicator or index. It is easy to define, but hard to understand its overall characteristics (Kim 2006). Traditionally, friction is defined as the force resistance to relative motion developed between two bodies in contact with one another, or more comprehensively, as the latent resistive force which opposes incipient movement at and parallel to the slip plane of two interfacing surfaces (Ludema 1987).

Figure 3.1 illustrates a friction event between a shoe heel and a floor surface during dynamic slip resistance measurements. As shown in Fig. 3.1a, if no adhesion takes place, then the only alternative interaction which would result in a resistance to motion would be one in which material is deformed and displaced during relative motion. It is needed to consider two interactions of this type: microscopic and macroscopic interaction (Moore 1975).

The microscopic interaction illustrated in Fig. 3.1a requires deformation and displacement of the interlocking surface asperities. And the macroscopic interaction illustrated in Fig. 3.1b means that the asperities of the harder material plough groove in the surface of the softer.

Characterization of the relative magnitude of this force is done most conveniently by expressing it as a ratio of the force required to just overcome the resisting force to the force acting perpendicular to the two surfaces in contact. That is, a COF is a property of the two interfacing and interacting surfaces and serves as a measure of their micro- and macro-roughness, inter- and intra-molecular forces of attraction and repulsion, and their viscoelastic (polymer and/or elastomer deformation) properties (Miller 1983).

As such, surface topographies of both bodies, contact areas, contact time, the velocity of movement, pressure, material types, and environments are major contributing factors to the COF results. The COF is also referred as either static coefficient of friction (SFC or SCOF) or dynamic (or kinetic) coefficient of friction (DCOF or DFC) depending on whether it is a measure of the forces at the instant relative motion begins or after there is a continuous, uniform sliding motion, respectively (Moore 1972, 1975).

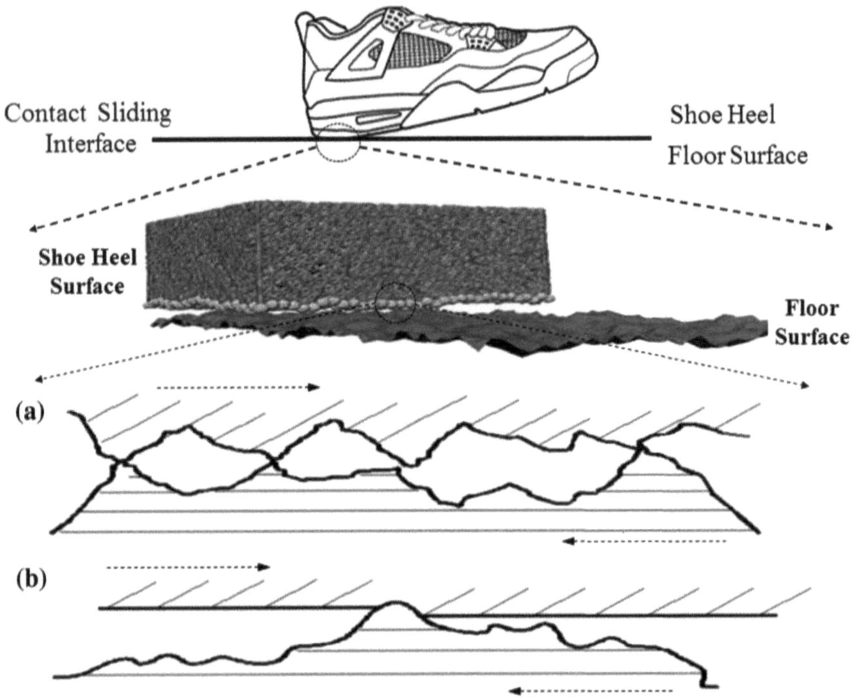

Fig. 3.1 Schematic illustrations of a friction event between a shoe heel and floor surface during dynamic slip resistance measurements: **a** asperity interaction only and **b** microscopic interaction, respectively

Because of the complex nature and multi-factors involved, the measured COF quantities seem to show inconsistencies even as the same shoe-floor combinations are employed for assessing their slip resistance performance. This reality has been recognised as a great concern when different friction testers, sensors, and/or protocols are adopted to measure slip resistance performance in the recent literature (Kim 2004a, 2006, 2015a; Kim and Nagata 2008a, b). However, most slip and fall safety assessments have simply reported that a specific shoe or floor surface resists the movement of a particular floor surface or one's shoe sole/heel across its surface. Wherein co-equal contributions of the footwear or floor surface are either ignored or not even considered in most slip resistance measurements.

As addressed in the above, the COF quantity is not a constant for any particular material but is typical of two bodies sliding against each other under a given set of environmental conditions. Therefore, either of the following questions would be an unreasonable enquiry (Kim and Nagata 2008a, b; Kim 2006, 2015a):

(1) What is a COF of this floor or floor covering? or
(2) What is a COF of that shoe?

The question should be raised by either:

(1) What type of shoes was measured or tested against which floor surfaces? or
(2) What is a COF index or quantity between this shoe and that floor surface?

It also should be noted that most slip resistance measurements have simply reported as a routine test such as started from clean and dry surface conditions and then moved into lubricated ones without any specific analysis of the surface conditions of tested shoes and floors under the clean and dry ones. However, such routine practices seem to oversimplify critical issues on slip resistance properties between the shoe and floor.

For example, surface circumstances of the rubbed (or tested) shoes and floors would be substantially from their initial ones. This means that the early surface finishes of shoes and floors were worn and changed in a way during dry sliding friction measurements. As a result, topographic features of both bodies would have different conditions from their original ones as the results of wear developments. Under such conditions, the following fundamental questions are specifically challenged for the assessments of slip resistance performance:

(1) How will the modified surfaces of the shoe and floor affect the slip resistance performance in the short- and long-term conditions?
(2) How will the worn (or tested) shoes and floors affect the slip resistance performance under lubricated conditions?
(3) Are there any differences in slip resistance performance between the worn shoes and a new floor surface as comparing the slip resistance result between the worn shoe and floor?
(4) Are there any differences in slip resistance performance between the worn floor and a new shoe as comparing the slip resistance result between the worn shoe and floor?

3.6 Friction Development Between Two Solid Surfaces

Friction has been studied since the early investigations of Leonardo da Vinci, Amontons, Coulomb, and Euler, and experimental investigations were used to measure the friction force between contact surfaces (Dowson 1979; Czichos 1986). In the late 14th Century, the "classical laws" of friction were first stated by Leonardo da Vinci. In 1699, Amontons rediscovered these laws and confirmed by experiments that the friction force was independent of the area of the surfaces and was directly proportional to the normal load as shown by Eq. (3.1).

$$F_S = \mu \times F_N \tag{3.1}$$

where $\boldsymbol{F_S}$ is the sliding traction force, $\boldsymbol{F_N}$ is the normal force, and $\boldsymbol{\mu}$ is the apparent coefficient of friction.

In 1785, Coulomb confirmed the first two laws of the classical laws of friction and added the third law that the friction force was independent of the velocity of sliding (Bowden and Leben 1939; Brungraber 1976; Tabor 1981). In 1835, Morin introduced the idea that friction could be differentiated into static and dynamic friction (Dowson 1997) Static friction was considered the force necessary to begin motion rather than maintain it whereas dynamic friction was usually slightly smaller and was described as the force necessary to maintain the motion (James 1983). However, in the case of non-homogenous and different surfaces, all of these laws were violated (Derieux 1934).

Despite considerable extents of experimental and analytical research, no 'simple' theoretical model has been developed to calculate the friction between two given surfaces (Suh and Sin 1981; Czichos 1986). It was stated by Bowden and Leben (1939) that there is

> … no clear understanding of the mechanism of friction of sliding solids (p. 371).

Our knowledge of the sliding process is considered inadequate (Rabinowicz 1956, 1958) and friction still remains one of the most familiar and yet least understood facets of mechanics (Brungraber 1976). A theoretical explanation of friction still does not exist and there is no simple model that allows the prediction or calculation of friction for a given pair of solid surfaces (Czichos 1986). In addition, Heilmann and Rigney (1981) stated that

> Most of us can agree on a simple definition of friction, but when we try to go much further, we find considerable disagreement about its fundamental nature. Moreover, none of the analytic expressions which have been developed can be used to calculate reliable values for friction coefficients (p. 15)

The current state of our knowledge on the friction mechanism is covered below and will be extended to cover surfaces that are not sensibly solid but are granular instead (Barry and Milburn 2015). When Coulomb established the 'classical laws' of friction, the physical model of the process he attributed friction to the force required to slide rigid high asperities on a rough surface over the asperities of the other surface by a lifting action. Coulomb also recognised that adhesion played a role, but it was Desagulier in 1854 who first suggested that adhesion was an important factor in friction (Johnson 1981).

The idea of surfaces with high friction being rough and surfaces with low friction being smooth originated with Coulomb. The apparent correlation between rough and smooth surfaces, however, is now considered to be false as expressed clearly by Tabor (1981).

> … it is still customary to refer to frictional surfaces as rough, and frictionless surfaces as smooth - this is spite of the fact we know that this correlation is false and completely out of date (p. 170)

The Coulomb laws of dry friction relevant to the understanding of traction mechanism are:

(1) The friction force is dependent on the kinds of materials in contact and their roughness;
(2) The static friction force developed at the point of sliding is dependent upon the normal force applied to the surfaces;
(3) The friction force is independent of the apparent geometric area of contact; and
(4) The friction force is inversely dependent on the velocity—decreasing as velocity increases (Czichos 1986; Nigg et al. 1986; Nigg and Segesser 1988; Jenson and Chenoweth 1990; Nigg 1990).

However, if the surfaces were contaminated, the corresponding friction laws become:

(1) The friction force is independent of the kinds of materials in contact and their roughness.
(2) The static friction force developed at the point of sliding is independent of the normal force applied to the surfaces.
(3) The friction force is dependent upon the apparent geometric area of contact.
(4) The friction force is dependent upon the velocity—increasing as velocity increases (Jenson and Chenoweth 1990).

Clearly, the laws for lubricated surfaces are different to those for dry friction, because of the effect of the lubricant films between the surfaces. In the case of shoe-floor friction system, the types of walkway surfaces are so diverse and walking environments encountered can be varied unpredictably such as dry, dusty, fibrous, granular, organic, wet, moist, soapy, or oily. For example, from the experiments conducted by Van Gheluwe et al. (1983) and Valiant (1994), the friction force for shoes with rubber or polymer outsoles moving over artificial turf surfaces increased with growths in contact area, which violated the 'classical' laws of friction. Bonstingl et al. (1975) also reported this finding.

However, for dry sliding friction, the friction force was independent of the area of contact (Jenson and Chenoweth 1990) contrary to the above findings. A notable statement that summarised the current view on the nature of friction was given by Heilmann and Rigney (1981).

> Two major approaches have dominated attempts to understand the origin of sliding friction. One of these has emphasised the role of surface roughness and the interlocking of surface asperities of various geometries. The other has focused on the role of adhesion and direct inter-atomic forces.
>
> …
>
> The most widely accepted model for friction, that of Bowden and Tabor, incorporates both of these approaches since it involves adhesion of surface asperities to form junctions and subsequent shear of junction materials. In this model one assumes homogeneous and isotropic materials which are in contact only at surface asperities. These asperities deform elastically or plastically until the contact area, $\boldsymbol{A}$, supports the normal load, $\boldsymbol{L}$. (p. 15)

Therefore, analytical investigations of friction behaviours currently attribute friction to a complex molecular-mechanical interaction that occurs between the contacting surfaces (Barry and Milburn 2015; Kim 2016). This complex interaction was assumed to be due to the combined effects of asperity deformation, ploughing by hard surface asperities and wear particles, and adhesion between flat surfaces (Suh and Sin 1981; Czichos 1986).

Three basic notions are used in an attempt to explain the sliding process and how friction develops between unlubricated sliding solids (Tabor 1981; Czichos 1986). To understand the sliding process and how friction is developed requires an understanding of:

(1) the area of real or true contact between the sliding surfaces;
(2) the type of strength of bond (adhesion) that is formed at the interface where contact occurs; and
(3) the way in which the material in and around the contacting region is sheared and ruptured during sliding (by ploughing or deforming).

Firstly, the real area of contact (RCA) is the sum of each individual contact at the asperities. This depends on the topography of each surface (shape, number, height, and distribution of the asperities) and their material deformation properties such as whether they are elastic, plastic, viscoelastic, viscoplastic, brittle, or combinations of these; and the value of the following variables: Young's modulus, yield pressure, plastic index, and hysteresis loss leading to a time dependence.

As sliding occurs between the surfaces, the RCA varies in value and at present, it is impossible to experimentally measure. It can only be deduced before and after the surface contact (Tabor 1981; Czichos 1986; Lambe and Whitman 1979; Chang et al. 2001). At the molecular level, smooth solid surfaces, in fact, are not perfectly smooth. They have valleys and ridges or asperities, and at a given instant, some of these asperities will be touching, as shown in Fig. 3.1. The sum of these contact areas is the RCA and is usually smaller than the apparent geometric area of contact (ACA). Topographies of the surface roughness are dependent on the distribution of asperities and the deviation of their heights and their slopes (Czichos 1986).

Secondly, the adhesion component of friction or the strength bond arises from the adhesion of the molecules in the surfaces in contact with each other. This adhesion may be so strong that at some contacts, tiny fragments are torn off and stick to the other (Rabinowicz 1956). The larger and short-range forces, where the surfaces are in more intimate contact, include metallic, covalent, and ionic forces, and the smaller and longer-range forces are termed as van der Waals bonds (Czichos 1986; Chang et al. 2001).

Of significance to understanding the shoe-floor interaction is the point made by Tabor (1981) that the bond could develop in a region of failure distant from the stronger surface interface.

> It [the bond] may depend to some extent on the mutual orientation of the two surfaces but broadly speaking the bond will be as strong as the metal itself. In fact, because of the plastic yielding and work hardening, the interface may be stronger than the undeformed material in the hinterland. This means in sliding, separation will not occur in the interface itself but at some distance away (Tabor 1981, p. 173).

Thirdly, the material behavioural components of friction involve two mechanisms: the ploughing component and the deformation component, respectively. In contacts where one surface is harder than the other, its asperities and wear particles penetrate into the softer surface and plough through it. The deformation component depends on the surface material load-deformation properties and slip-line theory. That is, the materials' Young's moduli (***E***), hardness (***H***), and shear strengths, and the use of rigid perfectly plastic behaviour.

How the asperities in contact respond to each other also depends on their material deformation properties, that is, their Young's moduli, and hardness, along with the yield pressure (***py***). For inelastic materials like polymers, viscoelastic, and viscoplastic, relaxation effects lead to a time-dependence of the contact area and hysteresis losses associated with the loading-unloading cycles. However, the ploughing and deformation components overlap separately and interact in a complex way in reality (Czichos 1986).

Whilst there is no accepted theory of friction, the complex molecular-mechanical interaction models mentioned above have been used to explain how the friction force is developed. Clearly, the types of materials in contact and their geometry structures shave an important influence on the friction force. In the case of footwear, the materials in contact would be shoe outsoles/heels made from polymers, plastics or rubber with various patterns or cleat configurations, interacting with either natural or artificial floors and walkways under a range of indoor or outdoor environments. However, it is very clear that the mechanisms explored to explain dry friction behaviours can provide the basis to explain the mechanisms associated with the shoe-floor interaction (Barry and Milburn 2015).

3.7 What Does a COF Quantity Mean?—Misconception and Restraint

The answer seems to depend on who is asked and what measuring devices are used (Kim 2006). For those are concerned with the definition as it applies to slip resistance between the footwear and underfoot surface, there are many definitions from which to choose. However, all those definitions are simple and direct If a floor surface meets an SCOF and/or DCOF value of 0.4 or 0.5 or greater as measured by various types of slip testers, the flooring materials and products can be classified as "very slip resistant" (Strandberg 1983; Grönqvist et al. 1989, 2003; Redfern and Bidanda 1994; Fong et al. 2009; Nagata et al. 2009).

With this classifying guidance, however, the minimum COF value of 0.4 or 0.5 varies considerably from time to time, causing a great deal of confusion as a safety threshold for the footwear and flooring manufacturers and consumers. A major confusion and critic on the COFs seem to start at this point. Since the reference value of 0.4 or 0.5 was established, many different types of testers for the slip resistance measurements have been developed to claim the measure of SCOF as well as DCOF as their safety threshold.

However, what an important point to note is that not all the testers would give the same result and should not be used as a basis for comparison by other test devices. That is, because each tester and/or apparatus have different theoretical and mechanical principles, it seems to be unreasonable to adopt the reference value without any citation to the instrument used for slip resistance measurements.

It is also reported that the COFs are primarily intended for scientific research, more sophisticated users and for evaluating brand new products (Miller 1983). That is, most of the measuring references established are used to focus on testing brand new products for shoes and flooring material rather than considering "real world service situations" . Those conditions should include surface degradation of both footwear and floors caused by wear progress, maintenance issues, and surrounding environments.

In particular, there is no existing source of definitive and quantified wear data for slip resistance measurements. Many round-robin comparisons of different instruments have been carried out, but there are still no reliable reference data to point towards the testers that most closely predict in-service performance. Therefore, these matters also need to be fully explored for the development of further effective concepts for the slip resistance measurement.

3.8 A Concept of Average COF—Case Study No. 1

As clearly discussed in the above Sects. 3.4–3.6, the key challenges identified from the currently practised slip resistance measurements are an oversimplified concept on friction behaviours and misunderstanding on the complex nature of tribo-physical characteristics of the footwear and walkway surfaces and their interactions. This means that an average or median COF reading is likely to be a quite improbable way to assess slip resistance properties between the shoe and floor.

Figure 3.2 shows an example of dynamic friction measurement results between a PVC shoe and a smooth vinyl floor as a function of test times under clean and dry surface environmental conditions. As shown in Fig. 3.1, the DFCs were gradually increased from 0.821 to 1.070 after 50 times of rubbings.

For a detailed investigation, the DFC results were divided into five groups. Each group was counted by ten times of rubbings. When the DFC results were compared between the group Nos. 1 and 5, there were large differences in their slip resistance performance despite a limited of rubbings. From the DFC results, the following fundamental questions are raised:

(1) How can define a DFC quantity (or value) between the PVC shoe and vinyl floor?

As shown in Table 3.1, the statistical summary of DFC results shows clear comparisons in each test group. Under this condition, it can be questioned what DFC quantity should be used for rating slip resistance performance between the PVC shoe and vinyl floor.

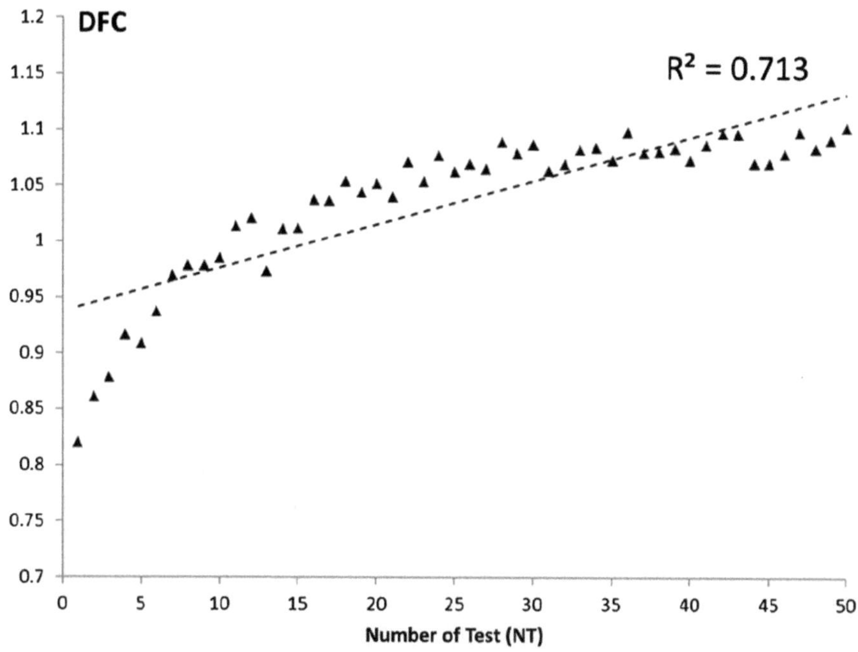

Fig. 3.2 A schematic plot of dynamic friction coefficients between a PVC shoe and a smooth vinyl flooring specimen under the dry conditions (Kim and Smith 2003)

Table 3.1 Basic statistical descriptions on the DFC results between a PVC shoe and a smooth vinyl floor after 50 times of dynamic friction tests

Basic statistics	DFC results				
	Group 1	Group 2	Group 3	Group 4	Group 5
Minimum	0.821	0.973	1.040	1.063	1.070
Maximum	0.985	1.054	1.089	1.098	1.102
Mean	0.924	1.025	1.069	1.079	1.087
Standard deviation	0.057	0.024	0.015	0.010	0.012
R-square	0.935	0.570	0.585	0.138	0.037

(2) How many test numbers are required for a valid DFC quantity?

Because the DFCs between the PVC shoe and vinyl floor continuously changed during the slip tests and were highly depended upon the numbers of rubbings, it is hard to reach a decision on a test number for attaining a reliable test result.

(3) What is a representing DFC quantity?

As shown in the statistical results in Table 3.1, it is also strongly questioned on a representative DFC quantity (or value) for this case. Should we choose 0.924

(shaded area) from the mean of group 1 or 1.087 (shaded area) from the mean of stage 5 or an average (1.006) of stages 1 and 5 or need more tests?

The above-raised questions are not only a matter of the current dilemmas on slip resistance measurements, but also a matter of primal issues for the requirement of in-depth understanding of friction and wear behaviours of the shoes and floors, and their interactive effects on slip resistance performance. Therefore, it is urged that the simplified concept of average or median COF readings for slip resistance assessments needs thorough investigations.

3.9 A Concept of Average COF—Case Study No. 2

The dynamic friction test performed in Fig. 3.2 was extended to further detailed investigations on the mis-leaded concept of average COF readings. The tests were performed 700 times of rubbings between the PVC shoe and smooth vinyl floor. Figure 3.3 plots the test results of expanded slip resistance measurements between them. It visibly shows that the DFCs continuously increased with additional rubbings.

In a very early stage of the dynamic friction measurements (until 50 times of rubbings), the DFC increased more rapidly as shown in Fig. 3.3b. As a result, massive topographic changes due to friction-induced wear developments were actively occurred at the surfaces of both shoe and floor and at the sliding interface between them. This was result clearly identified by the observation of extensive damages of the outmost layers of the heel surface due to continuous abrasion. This aspect was also easily confirmed by simple observations of large volumes of wear particles from the heel surfaces.

With further rubbings, the DFCs gradually increased until around 460 rubbings and became stabilised afterwards. However, after 600 times of rubbings, the DFC slightly increased again and then settled down. It was considered that the DFC results were directly related to wear behaviours between the shoe heel and flooring specimen. Because the shoe heel was a softer material than the floor, wear development was mainly concentrated on the heel surface. Thus, the shoe surface was continuously and massively changed by abrasive and fatigue wear. At the same time, it was also assumed that the floor surface was modified through various mechanisms such as material transfer, partial wear, and surface changes as found in the recent studies (Kim and Smith 2000; Kim 2015b, d, 2016).

The geometrical features of surface alterations and wear evolutions of both bodies were quantified by measuring surface roughness with two peak roughness parameters, $\boldsymbol{R_{pm}}$ and $\boldsymbol{R_{vm}}$ before and after the test. Table 3.2 summarises the result of two roughness parameters for the PVC shoe and vinyl floor. As shown in Table 3.2, the roughness parameters of two bodies showed clear changes after the whole test. In the case of the PVC shoe, the $\boldsymbol{R_{pm}}$ was significantly increased (77.5%) whilst the $\boldsymbol{R_{vm}}$ was largely decreased (45.9%). This result demonstrated that ploughing wear was a major cause of the changes of both peak roughness parameters.

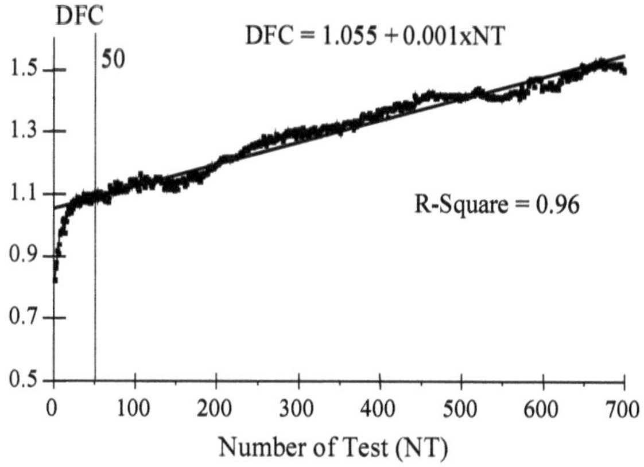

(a) Overall DFC results

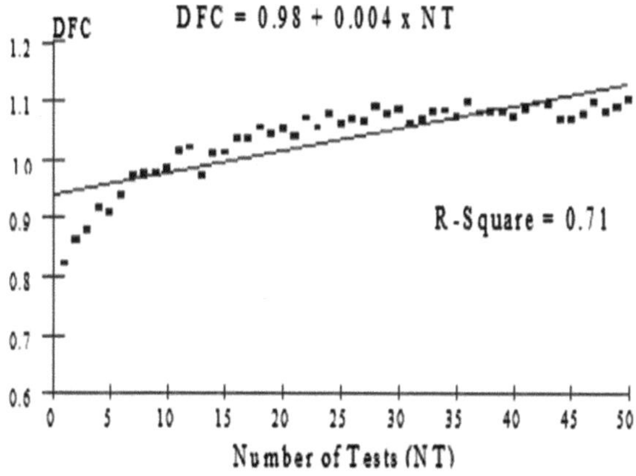

(b) DFC results of the initial 50 times of sliding

Fig. 3.3 A result of dynamic friction tests between a PVC shoe and a vinyl floor under the dry conditions: **a** overall DFC results and **b** DFC results of the initial 50 times of rubbings, respectively (Kim and Smith 2003)

Table 3.2 Comparison of the two extreme roughness parameters of the PVC shoe and the vinyl floor specimen before and after the tests

Material type	Test number	Surface parameters (μm)	
		R_{pm}	R_{vm}
PVC shoe	Initial	3.845	−5.145
	After 700	6.824	−2.783
Vinyl floor surface	Initial	8.968	−9.483
	After 700	5.137	−3.396

In the case of the floor counterface, however, both roughness parameter, $\boldsymbol{R_{pm}}$ and $\boldsymbol{R_{vm}}$, were largely decreased by more than 40 and 60%, respectively. The large decrease in the $\boldsymbol{R_{pm}}$ parameter seemed to lead a corresponding reduction in the adhesive force between the two mating bodies and thus, affect the slip resistance results. If the dynamic friction tests proceeded further, it would be most likely that the surface topography of the vinyl floor would become smoother and/or the surface asperities would become less sharp. The $\boldsymbol{R_{vm}}$ parameter also showed massive changes after the dynamic friction test. It could be assumed that wear products of polymeric particles from the shoe heel were embedded into the valley areas of the vinyl counterface. As a result, the asperity depth was largely reduced and caused the negative increase of the surface roughness parameter, $\boldsymbol{R_{vm}}$.

From the above test findings with measurements results of the two peak roughness parameters: $\boldsymbol{R_{pm}}$ and $\boldsymbol{R_{vm}}$, it becomes clear that asperity heights of the floor surface are also considerably affected by sliding friction events during the dynamic friction measurements. Hence, it should be pointed out that topographic changes of both shoe and floor surfaces should be measured and monitored regularly.

3.10 Issues of Frictional Force and Heel Strike Angle

3.10.1 Frictional Force

Figure 3.4 shows the displacement results of the frictional force between the PVC shoe and the vinyl floor specimen as a function of the test time during dynamic friction measurements. For the in-depth investigation on the frictional force component, the test results were plotted against the function of eight test numbers: 1, 15, 50, 80, 100, 300, 500, and 700, respectively. Then, the test results were divided into two groups for the detailed analysis: (1) below 100 times and (2) over 100 times of rubbings, respectively. This distribution was intended to observe major trends of frictional forces against a time interval (see, Table 3.3). The changes of frictional force showed similar pattern after 100 times of rubbings. From Fig. 3.4, the following important aspects were observed:

(1) An initial heel strike occurred at 0.8 s (Test No. 1).
(2) After the initial 50 times of rubbings, the heel-strike occurred at 0.7 s which was shortened by 0.1 s afterwards the initial heel strike. At this level, the frictional force was reached the second highest point (474 N) and the DFC value was also significantly increased (see, Fig. 3.3).
(3) Between 50 and 80 times of the rubbing numbers, there were no significant changes in the frictional force and the heel contact time.
(4) After 80 times of rubbings, the frictional force was drastically reduced (from 464 to 376 N) and the heel strike time was shortened again (0.06 s).

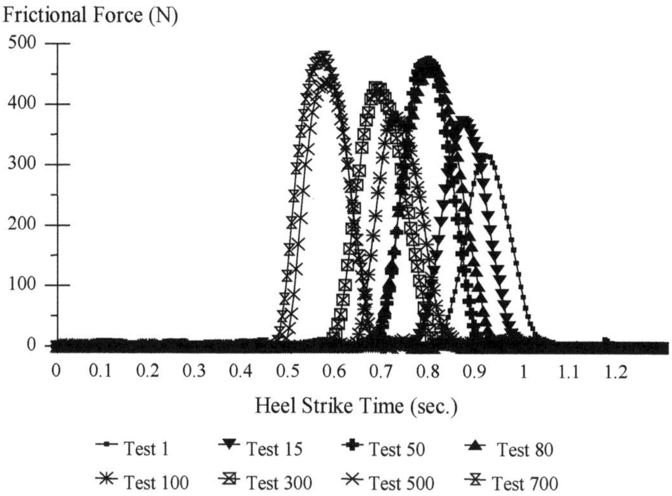

Fig. 3.4 A schematic plot of the changes in the frictional force component against a heel contact time interval as the function of test numbers (Kim and Smith 2003)

(5) At 100 times of rubbings, the frictional force and the DFC showed almost identical values as compared with the case of 15 times of sliding even though the heel strike time was largely shortened (over 0.1 s) (see, the shaded areas in Table 3.3).

(6) After 100 times of rubbings, the frictional force was gradually increased again and the DFC value was increased as well. This increase continued until the end of the test.

(7) The heel strike time between the 100 and 300 times of rubbings was reduced 0.1 s again and the heel strike occurred at 0.6 s.

(8) After 300 times of rubbings, the frictional force was not much increased until 500 times of sliding, but the DFC value was slightly increased and the heel strike time was reduced again and occurred at 0.5 s.

(9) After 500 times of rubbings, the frictional force and DFC value were slightly increased and showed the highest values respectively (see, Fig. 3.3 and the dotted area in Table 3.3). The heel strike time was just marginally reduced and occurred at 0.48 s.

From the above two groups of tests, it was found that the heel strike times were significantly reduced (over 40%) from the initial 0.8 s to the final 0.48 ones. It was considered that this seemed to be directly related to the growth of the heel contact areas. This means that the edge part of the heel area contacted against the floor surface is increased by wear progression alongside the sliding direction As a result, the heel strike times gradually reduced with further repeated rubbings.

Table 3.3 Comparison of the DFC values and frictional forces between the PVC shoe and the vinyl floor specimen during the dynamic friction tests

Test group	Test number	DFC	Frictional force (N)
Group I	1	0.821	284.9
	15	1.069	375.4
	50	1.344	473.8
	80	1.341	463.6
Group II	100	1.063	375.9
	300	1.215	429.0
	500	1.296	448.3
	700	1.395	492.9

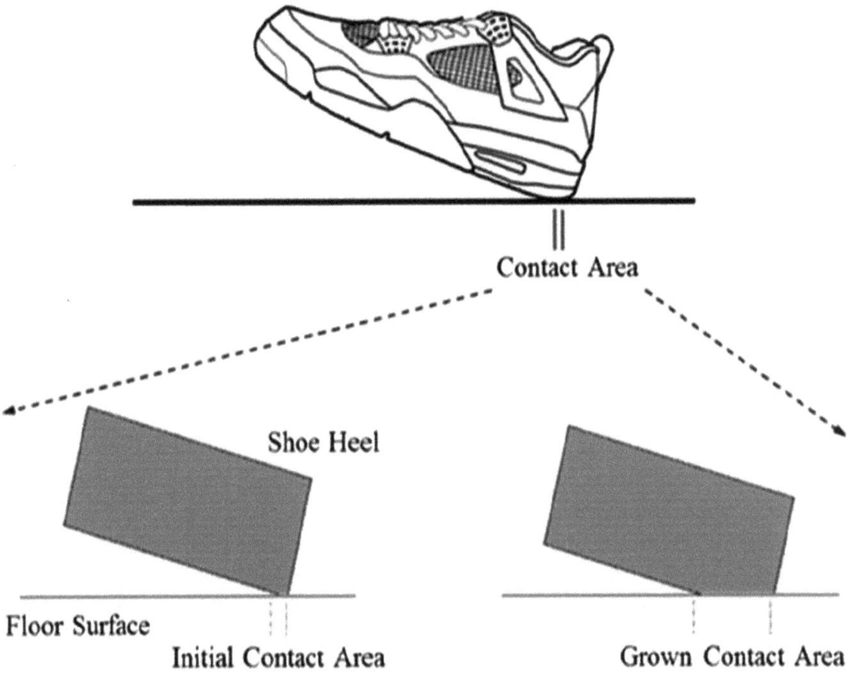

Fig. 3.5 Schematic illustrations of the variation of the contact area between the PVC shoe and the vinyl floor specimen after the dynamic friction tests (Kim and Smith 2003)

Figure 3.5 illustrates the above demonstrations on the possible growth of the heel contact area as a result of continual rubbings during the dynamic friction measurements. Changes on the heel surface were clearly observed at the end of the test. In this process, it can be certain that there is strong evidence of wear development at the sliding interface between the shoe heel and floor surface.

3.10.2 Heel Strike Angle

Figure 3.6 shows the inclusive changes of the heel strike angle during the test. The heel strike angle was plotted against the function of eight test numbers: 1, 15, 50, 80, 100, 300, 500, and 700, respectively. From this figure, the following important aspects were found:

(1) The initial heel strike angle was very unstable. At the moment of the shoe heel initially contacted against the floor surface, three random spikes were observed at 0.27, 0.48, and 0.75 s respectively. These three spikes seemed to demonstrate the possible initiation of wear developments between the PVC shoe and vinyl floor. This means that the heel surface started to build up massive amounts of pressures caused by sliding friction as observed in a previous study (Kim et al. 2001). During this process, the displacement of the heel strike angle seemed to become unsettled.
(2) With further rubbings, the heel strike angle was stabilised and the heel striking time was shortened as well. As a result, the contact area of the heel surface became wider and smoother. This result indicated that progressive wear behaviours between the two surfaces gradually contributed to the changes of the surface conditions of both bodies.
(3) The heel toe-off time was also shortened with the continuous rubbings so that the slope line of each angle displacement was gradually increased.

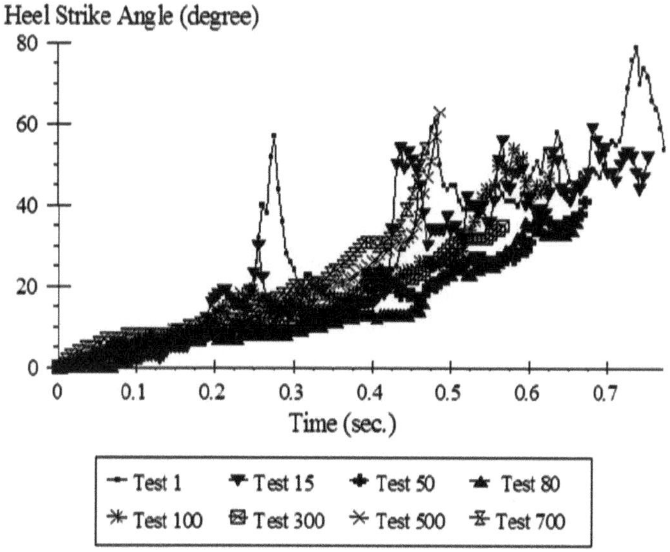

Fig. 3.6 A schematic plot of the changes in the heel strikes angle against a time interval as the function of test numbers (Kim and Smith 2003)

The above findings unambiguously demonstrated the importance of tribo-physical behaviours on the slip resistance performance between the shoe heel and floor surface. For example, changes of the frictional force showed clear evidence on the wear episode of both bodies of the shoe and floor even though all the experimental conditions such as the shoe and floor specimens and the vertical load were kept constant. From these results, the following important tribological aspects between the shoe and floor are found:

(1) The initial sliding stage (up to 50 times of rubbings) seems to be one of the most important moments for slip resistance measurements between the shoe heel and floor surface. During this early period of friction measurements, almost all the major tribolo-physical characteristics such as surface alterations, physical, mechanical, and chemical interactions, and wear growths seemed to occur on the surfaces of both shoe and floor. Hence, the number of test rubbings should be carefully considered when slip resistance measurements are conducted and reported.
(2) Changes of the frictional force were a serious issue that affects the slip resistance properties because its variation directly involved the DFC results.
(3) The initial heel strike time became shorter due to an increase of the contact area of the shoe heel. This outcome also influenced the DFC results.
(4) The initial heel strike angle was gradually increased so this consequence accordingly affected the DFC results.
(5) The surface topographies of both shoe and floor were significantly changed by friction caused wear developments. This wear effect on slip resistance performance should be thoroughly investigated with friction measurements.

3.11 Maintenance Issues

Slip resistance properties of textured floors and floor coverings may quickly lose without effective cleaning procedures and maintenance. Floors and floor coverings themselves significantly influence the degree of slip resistance (Murphy 2003). For example, soft flooring materials such as rubber or urethane are less slippery than hard surfaces such as ceramics or metals. Unfortunately, however, some of the softer materials are not viable for industrial applications. These surfaces generally require abrasion resistant functions that can withstand heavy impacts and traffic conditions (Murphy 2003). To accommodate these conditions, most slip-resistant floors for the industrial applications are composed of a polymeric material (Shore Hardness D > 50) which is usually incorporating some degree of aggregate (ASTM D2240 2000).

The porosity of floor surfaces would affect slip resistance properties under lubricated environments. For example, concrete slabs absorb waters, whilst waters stay on the surface if it is coated with epoxy. To address this issue, polymeric floors are installed with various scales of textured surface finishes. This allows waters (or other contaminants) to move out of the way of contacting surfaces. The surface topographies or finishes can be modified by applying a stipple finished polymer to an aggressive texture using bonded aggregate (Murphy 2003). Most industrial environments require only a slight non-skid texture to avoid slip and fall incidences. This can be accomplished using a fine aggregate incorporated into the topcoat surface of the floor.

Regarding the slip resistance properties for the floor surface, another important issue to consider is that the degree to surface topographies has a direct correlation to the ease of cleaning (Gould 2003). A high profiled or textured floor finish seems to be hard to clean and eventually lose its topographic features due to inability to remove the dirt and/or contaminants. Some cleaning materials also may leave residuals that can contribute to the slipperiness of the floor surface. It would be best to match the cleaner to the contaminant followed by effective removal of the detergent and contaminants.

Floors and flooring materials gradually wear and damage with use (Kim et al. 2013; Kim 2015b). For example, physical erosions of the floor would polish the surface finishes, increasing the risks of slipperiness. Depending upon the materials applied in the floors and floor coverings, this process can be extended greatly. Although surface features and finishes would definitely impact on the cleanability of surfaces, the effect on cleanability and maintenance issues has not yet been studied to any great extent. Therefore, researchers also should pay more attentions to those areas as well.

3.12 Conclusions

A simple format of friction measurement may misrepresent the complex nature of slip resistance properties amongst the shoes, floors, and environments. Facilitated routine friction measurements from laboratory environments also could oversimplify essential aspects of slip resistance properties. Although there has been considerable research progress on the understanding of slip resistance properties and its performance, it would be probably true to mention that none of the COF measurements reported to date could be regarded as final objective values for any chosen shoe-floor-environment combination.

As long as controversies around the friction measurement as a format of COF index remain, improvements in the principal concepts and methodologies on slip resistance determinations are urgently required. Therefore, it is essential for researchers working on slip and fall safety to find advanced concepts and measuring methods to assess the slipperiness for compensating the current limitations on slip resistance measurements. This should be based on thorough knowledge of the multifaceted nature of friction and wear behaviours and surface analysis techniques to explore the surface features of shoes and floors and their interactive effects on slip resistance performance.

Potential forms of friction and wear behaviours of the shoe and floor surface were clearly discussed by demonstrating the complexity of mechanics and mechanisms involved. The contribution of surface roughness to slip resistance was directly linked to the fundamental information for friction and wear behaviours of the shoes and floors. However, to date, these factors have hardly been considered in this research area. Therefore, this chapter was focused on broadening the knowledge base and developing new notions for characterising slip resistance properties from a tribological point of view on which advancements in the validity and reliability of slip resistance measurements might be made.

It is wished that suggestions from this chapter would not only provide a sound theoretical foundation for the understanding of both friction and wear phenomena between the footwear and underfoot surfaces, but also enhance the creditability of overall pedestrian fall safety assessments. Therefore, this chapter may provide a way of enhancing the reliability of fall safety determinations over the current methodology for COF readings.

3.13 Chapter Summary

This chapter strategically analysed tribo-physical characteristics of the frictional force component in the friction measurement caused by surface changes and wear progress that took place on the sliding interface between the PVC shoe and the vinyl floor specimen. Wear events with sliding friction significantly changed the topographic conditions of both surfaces with consequent effects on the frictional force.

With the immense surface alterations and wear developments, the frictional force was significantly influenced by the sliding system between the shoe heel and the floor surface even though all the experimental conditions were kept constant. This finding clearly indicates that the COF measure is not a constant for any pair of shoe-floor combinations, sliding against each other under a given set of surface and environmental conditions. That is, slip resistance property depends not just on the friction when the slip starts, but on how the friction varies as a slip progresses. Therefore, a simple friction measurement is not an ideal way to evaluate the slip resistance properties.

Furthermore, it was uncovered that friction was very sensitive to the state of pairing surfaces that were in contact. In this context, this chapter critically discussed fundamental but major issues on the measurements of slip resistance properties between the shoe and floor and addressed some critical problems in the interpretation of COF results, including its varying features during the dynamic friction measurements.

The new viewpoint explored in this chapter may provide a way of dealing with a reliability matter for the current concept on averaged COF readings as a measure for pedestrian safety thresholds and make some suggestions towards an account of the underlying tribological mechanisms. Therefore, this chapter principally focused on broadening the knowledge base and developing new ideas on which improvements in the validity and reliability of slip resistance measurements might be made.

References

ASTM D2240-64T. (2000). *Standard test method for rubber property—Durometer hardness*. ASTM Committee D11 on Rubber. Available at: http://www.abqindustrial.net/store/images/products/dmt/RX-DD/d2240.pdf

Barry, B., & Milburn, P. (2015). Tribology, friction and traction: Understanding shoe-surface interaction. Available at: http://www98.griffith.edu.au/dspace/bitstream/handle/10072/55722/87973_1.pdf?sequence=1

Bonstingl, R. W., Morehouse, C. A., & Niebel, B. W. (1975). Torques developed by different types of shoes on various playing surfaces. *Medicine & Science in Sports & Exercise, 7*(2), 127–131.

Bowden, F. P., & Leben, L. (1939). The nature of sliding and the analysis of friction. *Proceedings of the Royal Society, 169*, 371–391.

Brungraber, R. J. (1976). *An overview of floor slip-resistance research with annotated bibliography*. National Bureau of Standard, NBS Technical Note 895, U.S. Department of Commerce, National Technical Information Service, PB-248 985. Washington, January.

Chaffin, D. B., Woldstad, J. C., & Trujillo, A. (1992). Floor/shoe slip resistance measurement. *Journal of American Industrial Hygiene Association, 53*(5), 283–289.

Chang, W. R., Grönqvist, R., Leclercq, S., Myung, R., Makkonen, L., Strandberg, L., et al. (2001). The role of friction in the measurement of slipperiness, Part 1: Friction mechanisms and definition of test conditions. *Ergonomics, 44*(13), 1217–1232.

Czichos, H. (1986). Introduction to friction and wear. In K. Friedrich (Ed.), *Friction and wear of polymer composites*. Amsterdam, The Netherlands: Elsevier.

Derieux, J. B. (1934). The coefficient of friction of rubber. *Journal of the Elisha Mitchell Scientific Society. Rubber Chemistry and Technology, 8*, 441–442.

Dowson, D. (1979). *History of tribology*. London: Longman, UK.

Dowson, D. (1997). *History of tribology* (2nd ed.). Professional Engineering Publishing.

Fong, D. T. P., Hong, Y., & Li, J. X. (2009). Human walks carefully when the ground dynamic coefficient of friction drops below 0.41. *Safety Science, 47*(10), 1429–1433.

Gould, P. (2003). Smart, clean surfaces. *Materials Today, 6*(11), 44–48.

Grönqvist, R. (1995). Mechanisms of friction and assessment of slip resistance of new and used footwear soles on contaminated floors. *Ergonomics, 38*(2), 224–241.

Grönqvist, R., Hirvonen, M., Rajamäki, E., & Matz, S. (2003). The validity and reliability of a portable slip meter for determining floor slipperiness during simulated heel strike. *Accident Analysis & Prevention, 35*(2), 211–225.

Grönqvist, R., Roine, J., Järvinen, E., & Korhonen, E. (1989). An apparatus and a method for determining the slip resistance of shoes and floors by simulation of human foot motions. *Ergonomics, 32*(8), 979–995.

Harrison, R., & Malkin, P. (1983). On-site testing of shoe and floor combinations. *Ergonomics, 26*(1), 101–108.

Health and Safety Executive (HSE). (2011). *Investigation of slip resistance and the hygienic cleaning of floors in hospital settings*. RR889 research report. Available at: http://www.hse.gov.uk/research/rrpdf/rr889.pdf

Heilmann, P., & Rigney, D. A. (1981). Sliding friction of metals. In D. Dowson, C. M. Taylor, M. Godet, & D. Berthe (Eds.), *Friction and traction* (pp. 15–19). Guilford: IPC: Business Press.

Hoang, K., Stevenson, M. G., Nhieu, J., & Bunterngchit, Y. (1987). *Dynamic friction at heel strike between a range of protective footwear and non-slip floor surface*. Report No. CSS/1/87. Center for Safety Science, University of New South Wales, Australia. July.

Hoang, K., Stevenson, M. G., & Willgoss, R. (1985). Measurement of dynamic friction between shoe soles and walkway surfaces. In *Proceedings of the 22nd annual conference of the ergonomics society of ANZ* (pp. 265–171). Toowoomba. December.

James, D. I. (1983). Rubbers and plastics in shoes and flooring: The importance of kinetic friction. *Ergonomics, 26*(1), 83–99.

Jenson, A., & Chenoweth, H. H. (1990). *Applied engineering mechanics* (p. 164). Ohio: Glencoe/McGraw-Hill.

Johnson, K. L. (1981). Aspects of friction. In D. Dowson, C. M. Taylor, M. Godet, & D. Berthe (Eds.), *Friction and traction. Proceedings of the 7th leeds-lyons symposium on tribology*. The Institute of Tribology, Department of Mechanical Engineering.

Kim, I. J. (2004a). Development of a new analyzing model for quantifying pedestrian slip resistance characteristics: Part I—Basic concepts and theories. *Industrial Journal of Industrial Ergonomics, 33*(5), 395–401.

Kim, I. J. (2004b). Development of a new analyzing model for quantifying pedestrian slip resistance characteristics: Part II—Experiments and validations. *Industrial Journal of Industrial Ergonomics, 33*(5), 403–414.

Kim, I. J. (2006). The current hiatus in fall safety measures. In W. Karwowski (Ed.), *International encyclopedia of ergonomics and human factors-2005* (pp. 2572–2576). USA: Taylor & Francis Group, LLC.

Kim, I. J. (2015a). Practical design search for optimal floor surface finishes to prevent fall incidents, In B. Evans, (Ed.), *Accidental falls: Risk factors, prevention strategies and long-term outcomes* (pp. 80–103). Hauppauge, NY, USA: Nova Science Publishers, Inc. (Chapter 5).

Kim, I. J. (2015b). Slip-resistance measurements for assessing pedestrian falls: facts and fallacies. In B. Evans (Ed.), *Accidental falls: Risk factors, prevention strategies and long-term outcomes* (pp. 105–125). Hauppauge, NY, USA: Nova Science Publishers, Inc. (Chapter 6).

Kim, I. J. (2015c). Wear observation of shoe surfaces: Application for slip and fall safety assessments. *Tribology Transactions, 58*(3), 407–417.

Kim, I. J. (2015d). Research challenges on slip-resistance measurements for assessing pedestrian fall incidents. *Journal of Ergonomics*, 5(3). doi:10.4172/2165-7556.1000e142

Kim, I. J. (2016). Identifying shoe wear mechanisms and associated tribological characteristics: The importance for slip resistance evaluation. *Wear, 360–361,* 77–86.

Kim, I. J., Hsiao, H., & Simeonov, P. (2013). Functional levels of floor surface roughness for the prevention of slips and falls: Clean-and-dry and soapsuds-covered wet surfaces. *Applied Ergonomics, 44*(1), 58–64.

Kim, I. J., & Nagata, H. (2008a). Research on slip resistance measurements—A new challenge. *Industry Health, 46*(1), 66–76.

Kim, I. J., & Nagata, H. (2008b). Nature of the shoe wear: Its uniqueness, complexity, and effects on slip resistance properties. In *Contemporary Ergonomics 2008* (Vol. 15, pp. 728–734). Taylor & Francis.

Kim, I. J., & Smith, R. (1998a, July). *A study of the comparative geometry mating between the surfaces of the shoe and floor in pedestrian slip resistance measurements*. The 5th Pan-Pacific Conference on Occupational Ergonomics (pp. 34–37). Kitakyushu, Japan.

Kim, I. J., & Smith, R. (1998b, August). *Tribological characterization of the frictional force component in pedestrian slip resistance measurements*. Third World Congress of Biomechanics (WCB '98). Hokkaido, Japan.

Kim, I. J., & Smith, R. (2000). Observation of the floor surface topography changes in pedestrian slip resistance measurements. *Industrial Journal of Industrial Ergonomics, 26*(6), 581–601.

Kim, I. J., & Smith, R. (2003). A critical analysis of the relationship between shoe-floor wear and pedestrian/walkway slip resistance. In M. I. Marpet, & M. A. Sapienza (Eds.), *Metrology of pedestrian locomotion and slip resistance, American society of testing and materials*. Special technical publication (Vol. 1424, pp. 33–48). Philadelphia, USA: ASTM International.

Kim, I. J., Smith, R., & Nagata, H. (2001). Microscopic observations of the progressive wear on the shoe surfaces which affect the slip resistance characteristics. *Industrial Journal of Industrial Ergonomics, 28*(1), 17–29.

Lambe, T. W., & Whitman, R. V. (1979). Soil Mechanics, SI version. *Series in Soil Engineering*. New York, NY: John Wiley and Sons.

Layne, L. A., & Landen, D. D. (1997). A descriptive analysis of nonfatal occupational injuries to older workers, using a national probability sample of hospital emergency departments. *Journal of Occupational and Environmental Medicine, 39*(9), 855–865.

Leclercq, S., Tisserand, M., & Saulnier, H. (1993). Quantification of the slip resistance of metal surfaces at industrial sites. Part I. Implementation of a portable device. *Safety Science, 17*(1), 29–39.

Ludema, K. C. (1987). *Friction: A study in the prevention of seizure*. ASTM Standardization News. May 54–58.

Miller, J. M. (1983). Slipper work surfaces: Towards a performance definition and quantitative coefficient of friction criteria. *Journal of Safety Research, 14*(4), 145–158.

Moore, D. F. (1972). The friction and lubrication of elastomers. In *International series of monographs on materials science and technology* (Vol. 9). Headington Hill Hall, Oxford: Pergamon Press Ltd.

Moore, D. F. (1975). *Principles and application of tribology*. Oxford: Pergamon Press.

Murphy, T. (2003). Slip-resistant flooring (pp. 390–396). The Society for Protective Coatings.

Myung, R., Smith, J. L., & Leamon, T. B. (1992a). Slip distance for slip/fall studies. In *Advances in industrial ergonomics and safety* (Vol. IV, 983–987).

Myung, R., Smith, J. L., & Leamon, T. B. (1992b). Slip distance as an objective criterion to determine the dominant parameter between static and dynamic COFs. In *Proceedings of the human factors society, 36th annual meeting* (pp. 738–741).

Nagata, H., Watanabe, H., Inoue, Y., and Kim, I. J. (2009). Fall and validities of various methods to measure frictional properties of slippery floors covered with soapsuds. In *Proceedings of the 17th world congress on ergonomics*. Beijing, CD-ROM.

Nigg, B. M. (1990). The validity and relevance of tests used for the assessment of sports surfaces. *Medicine and Science in Sports and Exercise, 22*(1), 131–139.

Nigg, B. M., Bahlsen, A. H., Denoth, J., Luethi, S. M., & Stacoff, A. (1986). Factors influencing kinetic and kinematic variables in running. In B. M. Nigg (Ed.), *Biomechanics of 16 running shoes* (pp. 139–159). Champaign, IL: Human Kinetics.

Nigg, B. M., & Segesser, B. (1988). The influence of playing surfaces on the load on the locomotor system and on football and tennis injuries. *Sports Medicine, 5*(6), 375–385.

Perkins, P. J., & Wilson, M. P. (1983). Slip resistance testing of shoes—New development. *Ergonomics, 26*(1), 73–82.

Proctor, T. D., & Coleman, V. (1988). "Slipping and tripping accidents and falling accidents in Great Britain—Present and future. *Journal of Occupational Accidents, 9*(4), 269–285.

Rabinowicz, E. (1956). Stick and slip. *Scientific American, 194,* 109–118.

Rabinowicz, E. (1958). The intrinsic variables affecting the stick-slip process. *Proceedings of the Physical Society of London, 71,* 668–675.

Redfern, M. S., & Bidanda, B. (1994). Slip resistance of the shoe-floor interface under biomechanically relevant conditions. *Ergonomics, 37*(3), 511–524.

Scott, W. E. (2009). *Falls at work: protect your employees: statistics of accidents*. The National Safety Council.

Strandberg, L. (1983). On accident analysis and slip-resistance measurement. *Ergonomics, 26*(1), 11–32.

Strandberg, L., & Lanshammar, H. (1983). On the biomechanics of slipping accidents. In V. I. I. I. Biomechanics, H. Matsui, & K. Kobayashi (Eds.), *Biomechanics VIII* (pp. 397–402). USA: University Park Press, Baltimore.

Suh, N. P., & Sin, H. C. (1981). The genesis of friction. *Wear, 69*(1), 91–114.

Tabor, D. (1981). Friction-the present state of our understanding. *Journal of Lubrication Technology, 103,* 169–179.

Valiant, G. A. (1994). Evaluating outsole traction of footwear. In W. Herzog, B. M. Nigg & A. van den Bogert (Eds.), *Proceedings of the Eighth Biennial Conference* (pp. 326–327). Calgary, Canada: Canadian Society for Biomechanics.

Van Gheluwe, B., Deporte, E., & Hebbelinck, M. (1983). Frictional forces and torques of soccer shoes on artificial turf. In B. M. Nigg & B. A. Kerr (Eds.), *Biomechanical aspects of sports shoes and playing surfaces* (pp. 161–168). Calgary: University of Calgary Press.

Wilson, M. P. (1990). Development of SATRA slip test and tread pattern design guideline, slips, stumbles, and falls: Pedestrian footwear and surfaces. In B. E. Gray (Ed.), *ASTM STP 1103* (pp. 113–123). Philadelphia: American Society for Testing and Materials.

Yeoh, H. T., Lockhart, T. E., & Wu, X. (2013). Non-fatal occupational falls on the same level. *Ergonomics, 56*(2), 153–165.

Chapter 4
Tribological Approaches for the Pedestrian Safety Measurements and Assessments

4.1 Introduction

To effectively measure and assess slip resistance properties, it is crucial to fully understand the fundamental principles and underlying mechanisms of frictional behaviours and related tribo-physical characteristics that affect slip resistance performance between the walkway surfaces and footwear. Solid knowledge and comprehensive understanding of such causes and triggering factors would permit to make an improved analysis of slip resistance properties that contribute to or impede pedestrian fall safety and provide further reliable analyses and interpretations of measured results for slip resistance performance.

4.2 Tribo-Physical Approaches

4.2.1 Overview

It is frequently experienced that rough floors are more slip resistant than smooth ones, but there is a lack of scientific and engineering evidence to explain them. The very first investigation on surface roughness appears to be observed by a French engineer, Amontons in 1866, who realised that touching surfaces were not flat and asserted that surface roughness must enter each other (Kennaway 1970). Hunter (1930) reported that surfaces of the test floors and/or the surfaces of most shoe soles and/or heels became progressively smoother as tests were repeatedly carried out on the same spot and at the same time. According to Hunter, the static COF dropped 15–60% during the first 8–11 measurements.

Sigler et al. (1948) showed that good anti-slip properties under water wet condition were usually associated with rough particles that projected through water films and thus prevented its action as lubricants. Jung and Reidiger (1982) used a

I.-J. Kim, *Pedestrian Fall Safety Assessments*,
DOI 10.1007/978-3-319-56242-1_4

photo-optical method to measure the roughness of floor surfaces. They found that surface roughness of the floors was well related to subjective assessments of slipperiness.

> Jung and Reidiger (1982) also suggested that the mean peak-to-valley (R_{tm}) roughness parameter plus the mean sum of profile peaks (I) were the best objective measures for assessing surface roughness of the floors. They devised a classification for profiled and non-profiled floor surfaces and calculated drainage volumes, and volumes of liquid that could be held in the troughs below the raised profile. Recently, Nagata (1993) also pointed out that COFs were prone to be dispersive and changeable, depending upon factors such as contact areas and repetitions of measurements.

Despite the great efforts in many studies, there has been little success in solving the real-world problems of pedestrian slip resistance measurements. Amongst many possible reasons, ill-defined analyses seem to be mainly caused by oversimplified concepts on frictional phenomena between the shoe heels and floor surfaces. Hence, a fresh insight would be required to systematically approach multiple characteristics of friction and wear behaviours of the shoes and floors, their interactive tribo-physical characteristics, and overall effects on slip resistance performance (Kim 2016a, b). This approach would provide better-featuring slip resistance properties than a commonly practised mean or averaged COF reading. Whilst controversies around the friction measurement for slipperiness assessment remain (Chang et al. 2001; Kim 2004a, b, 2015a; Kim and Nagata 2008a, b), a tribological approach may provide an objective alternative to overcome the current limitations on slip-resistance assessments.

Therefore, this chapter robustly analyses the current concept on slip resistance measurements, which are based on simple COF quantity or index. With critical reviews and discussions on the current notion for slip resistance assessments, this chapter focuses on improving our understanding and building solid knowledge against the complex nature of frictional behaviours of the shoes and floors and their involved tribo-physical characteristics at the sliding interfaces between the shoe heels and floor surfaces. Accordingly, this idea and information may significantly improve the validity and reliability of slip resistance measurements which would provide better featuring slip resistance properties than the current concept which is commonly practised mean or averaged COF readings.

4.2.2 Limitations and Issues

The recent literature has clearly pointed out the importance of surface roughness to assessing both footwear and underfoot surfaces and their effects on slip resistance properties (Proctor and Coleman 1988; Manning et al. 1998; Harris and Shaw 1988; Kime 1991; Manning et al. 1990; Proctor 1993; Manning and Jones 1994; Grönqist 1995; Jones et al. 1995; Kim 1996a, b, c, d, 2004a, b, 2006a, b, 2015a, b, c, d, 2016a, b; Kim and Smith 1998b, 2000, 2001a, b, c, 2003; Chang 1998, 1999, 2000a, b; Chang and Matz 2000, 2001; Kim et al. 2001, 2013; Kim and Nagata 2008a, b).

But, the current research is still very limited to this area although this approach has shown promising results on identifying the roles of surface finishes of both shoes and floors on slip resistance performance. Hereafter, some of the recent studies that had dealt with surface roughness issues on slip resistance measurements are briefly reviewed and discussed to detect limitations and problems. Based on the reviews and discussions, further improvements are suggested for the future studies on fall safety assessments.

Braun and Roemer (1974) studied the effect of waxes on static and dynamic COFs and reported that slip resistance results were increased by an average of 50% after the floor surfaces were treated with polishing.

Bring (1982) reported the importance of surface roughness of floor and shoe materials regarding slip resistance properties. He suggested that surface roughness of the tested shoe and floor specimens should be stated with test methods for slip resistance measurements. For this purpose, a method for measuring the centre line average roughness parameter ($\boldsymbol{R_a}$) which was closely related to the corresponding International Organization for Standardization (ISO) system was proposed (ISO/R 468 1982).

Jung and Riediger (1982) used an inclined plane to determine static friction of 30 different floor coverings. They stated that gradings from surface roughness (peak-to-valley) reflected very well with subjective ratings whilst an angle of inclination was compared badly with subjective gradings.

Hoang et al. (1985) used a dynamic friction test rig to measure slip resistance performance between three flooring materials (cement, metal, and vinyl) and four shoes (leather, mica rubber, nitrile rubber, and PVC) under four different environments: dry, wet, soapy, and oily, respectively. They showed that a trend of slip resistance was increased with the floor surface roughness.

Wilson and Perkins (1985) reported an overall good agreement between a laboratory slip resistance test and a ramp test. They found out that the laboratory test was largely relevant to walking conditions.

Manning et al. (1985) tested different shoe materials and stated that the microcellular polyurethane (PU) was a better sole material for preventing slip accidents than the nitrile runner one under the mineral oil-covered environment.

Proctor and Coleman (1988) drew attention to the importance of surface roughness for the pedestrian fall safety problems. They compared COFs from various friction measuring devices and found out that hydrodynamic considerations were an important aspect in determining slip resistance performance between the footwear and floor surfaces against contaminated environments with different types of fluids.

With the hydrodynamic squeeze-film theory, Proctor and Coleman (1988) explained that a certain scale of surface roughness needed to improve slip resistance, depending on the viscosity levels of contaminants. They suggested that the details of surface roughness should be collected in any field comparison of test tools for slip resistance measurements.

Harris and Shaw (1988) assessed user's opinions of 10 contaminated floor surfaces with surface roughness measurements using the Suntronic 10 surface roughness meter. They reported a good correlation between the user's opinion ranking in slip resistance performance and surface roughness parameter, $\boldsymbol{R_{tm}}$ (the average maximum peak-to-valley height). They proposed that the $\boldsymbol{R_{tm}}$ of 8–10 μm was required to maintain a proper slip resistance under a wet environment.

However, it should be pointed out that this study oversimplified the important factors of surface features and their effects on slip resistance functioning. The following major questions are raised by the study of Harris and Shaw (1988):

(1) There was no clear explanation why they choose the peak-to-valley roughness parameter ($\boldsymbol{R_{tm}}$) amongst large numbers of surface roughness parameters for their study. Because this extreme parameter is a very sensitive indicator of high peaks or deep scratches in a surface (Sherrington and Smith 1987), the quantity of this high peak roughness parameter can be easily changed by friction processes. In addition, the $\boldsymbol{R_{tm}}$ parameter is calculated using data for the whole surface (assessment length) whereas only the highest points of the surface seem to actually contact with the shoe surface. Thus, frequent assessments of this parameter are required to correctly monitoring its changes during the slip resistance measurements.

(2) This study used a device that had a diamond-tipped stylus for the surface roughness measurements of floor specimens. Although this measuring device is generally well adapted for the measurements of surface texture, an avoidable limitation could be resulted from the size of the stylus. The combination of a finite tip radius (usually 2–15 μm) and an included angle prevented the stylus from penetrating fully into deep narrow features of the surface. In some applications, this can lead to significant measuring errors for the assessment of surface texture. Special styluses with chisel edges and minimum tip radii as small as 0.1 μm can be used to examine fine surface details where a conventional stylus would be too blunt, but all stylus methods inevitably produce some "smoothing" errors from the surface profiles due to the limited dimensions of the stylus tip.

(3) A further error can be introduced to examining compliant or smooth surfaces by stylus methods. The load on the stylus, although it may be small, seems to distort or damage the surface so non-mechanical contact method such as an optical interferometry is recommended for measuring surface roughness profiles (Bhushan et al. 1988). Although most of the profilometers provide good measurement results, another issue to consider to measuring surface roughness is that they have severe weaknesses in terms of detectability of detailed surface features and difficulties arising when three-dimensional data sets of surface features are required. Three-dimensional approaches provide more comprehensive information on the micro aspects of surface topographies the two-dimensional ones. This information would be a vital source for the investigation of friction and wear behaviours of both shoe and floor surfaces and their interactive effects on tribo-physical properties.

(4) Harris and Shaw reported uncommonly high readings from the measurements of surface roughness, but simply averaged all their data on surface roughness without any proper verification. However, it should be pointed out that tribo-physical aspects between the shoe heel and floor surface involve diverse and complex interactive modes. This aspect indicates that the surface finishes of both floor and rubber samples tested in Harris and Shaw's study seem to be largely modified by repetitive rubbings. The issue of topographic changes of both shoe and floor surface during slip resistance measurements are found in the literature (Kim 1996a, b, c, d, 2004a, b, 2015a, b, c, 2016a, b; Kim and Smith 1998b, 2000, 2001a, b, c; Chang 2000a, b; Kim et al. 2001, 2013; Kim and Nagata 2008a, b). Therefore, simple averaging of the surface roughness data without any systematic observation seems to be an inappropriate practice. Because each shoe-floor combination has a unique and multiple sets of frictional and wear features, it requires extra cares to interpret the measurement results of surface roughness for analysing slip resistance properties.

Stevenson et al. (1989) examined the effects of surface finishes for the commercially available floor surfaces against slip resistance properties. They showed that DFCs increased almost linearly with the centre line average (CLA, $\boldsymbol{R_a}$) roughness parameter of the floor surfaces. Under the oil-covered highly risky environment, there were large increases in the slip resistance performance as the floor surface roughness was increased.

On the other hand, further increments in the floor surface roughness did not make any difference against slip resistance performance. But, this study did not clearly demonstrate the effects of topographic features of floor and shoe specimens tested on slip resistance properties and discuss related tribo-physical characteristics at the sliding interfaces between the shoe and floor surfaces. The following key issues are questioned in the study of Stevenson et al. (1989):

(1) Dynamic friction test results between the concrete and metal floor specimens showed large differences in DFCs even though they showed similar scales of surface roughness. In the case of the concrete slab, the slip resistance property was largely improved as the surface roughness was increased from 8 to 28 μm (in $\boldsymbol{R_a}$ roughness parameter). Although the surface roughness of the concrete floor was increased up to 120 μm, there were no differences in DFCs. In contrast, the slip resistance results of the steel plates drastically increased under wet and soapy environments although the surface roughness was slightly increased from 1 to 4 μm in the *Ra* roughness parameter. However, there were no solid explanations on the above results of slip resistance properties.

(2) In the case of footwear, for example, the slip resistance performance of a Nitrile Rubber shoe disclosed better against the steel plate ($\boldsymbol{R_a}$ = 16 μm) than the concrete slab ($\boldsymbol{R_a}$ = 120 μm), which was approximately seven times rougher than the steel plate. This result showed that DFCs were not linearly increased with the surface roughness of floors. However, there was no clear explanation on this aspect of slip resistance result in the study.

(3) Polyurethane shoes were repeatedly rubbed against twelve different floor surfaces under five dissimilar environments. From this test circumstance, it could be anticipated that surface conditions of the shoe heels would be heavily modified by friction-induced abrasive wear developments. However, issues on topographic changes and wear development of the shoe surfaces were not explored at all in the study.

Manning et al. (1998) conducted a walking traction test to assess shoe slip resistance properties against a contaminated floor surface. They used different abrasive methods to simulate polishing effects of walking and reported a good correlation between the R_{tm} parameter of the shoe surfaces and measured friction results. However, it could be questioned that there would be big differences in topographic features of shoe specimens between the mildly worn heel surface by walking simulations and the heavily worn heel surfaces generated by the belt and orbital sanders. This aspect of topographic changes and wear developments on the shoe sole/heel surfaces was not investigated in this study as well.

Grönqvist et al. (1990) used different types of dynamic friction testers to simulate walking to measure slip resistance performance under contaminated environments. They reported a significant correlation between the measured mean COFs of 13 decks and underfoot surfaces on ships and the R_a roughness parameter of floor surfaces. This study also suggested that an adequate R_a value for the floor surface should be 7–9 μm for slip resistance under the glycerol-covered condition. However, they did not compare the changes of surface roughness characteristics before and after the tests. A study from Grönqvist et al. (1992) also reported a correlation between the COFs and the R_a roughness parameter of 19 different floor surfaces in the food industry. They reported that there was no significant correlation between them.

Chaffin et al. (1992) examined the effect of COFs on five variables such as a vertical load, velocity, shoe material, dry and wet conditions, and surface roughness. They concluded that the most significant effect on both static and dynamic COFs was roughness of the walking surfaces.

Proctor (1993) measured the surface roughness of 19 dissimilar floor surfaces and their COFs using three different types of slip testers such as the British ramp test, Tortus, and Pendulum tester. He found out that there was a significant relationship between the surface roughness and COF results from the ramp test. He also supported the requirement of minimum roughness for safe walking on wet floors, which was proposed by Harris and Shaw (1988).

4.2.3 Main Problems

As clearly discussed in the above, most studies in the literature show that there are common, but critical issues on the measurements of slip resistance properties, the analysis and the interpretation of slip resistance results. Such problems are briefly reviewed in the following:

Firstly, one of the most arguable issues is that commonly cited or referenced surface roughness parameters such as $\boldsymbol{R_a}$ and $\boldsymbol{R_{tm}}$ (or $\boldsymbol{R_t}$), are highly location-dependent and reveal limited surface features although they show fair correlations with frictional behaviours. For example, two sinusoidal surface profiles with identical amplitudes but different frequencies have the same scales of $\boldsymbol{R_a}$ and $\boldsymbol{R_{tm}}$ although their frictional behaviours seem to be quite different as shown in Fig. 4.1.

The $\boldsymbol{R_{tm}}$ parameter also has a limitation in representing the correlation between the slip resistance property and surface roughness. Because this parameter is calculated using roughness data for the whole surface profile, it seems to be difficult to distinguish what extent of the highest points of the floor surface may actually involve the contact with the shoe surface.

Secondly, another critical issue to consider is a method to calculating COFs. As clearly addressed the problems of average friction readings, this concept is still commonly practised in the measurements of slip resistance performance for the pedestrian fall safety assessment. This issue was also raised by the recent studies in the literature (Grönqvist 1995; Jones et al. 1995; Kim and Nagata 2008a, b; Kim 2015a). These studies pointed out the irrationality of average friction readings.

Grönqvist (1995) also mentioned that average friction readings were probably not at all decisive from the slip resistance point of view. He suggested that an instantaneous COF measure might be more relevant because the time available to achieve a sufficient COF to avoid slips and to counteract a possible fall were very short, only a few tenths of a second.

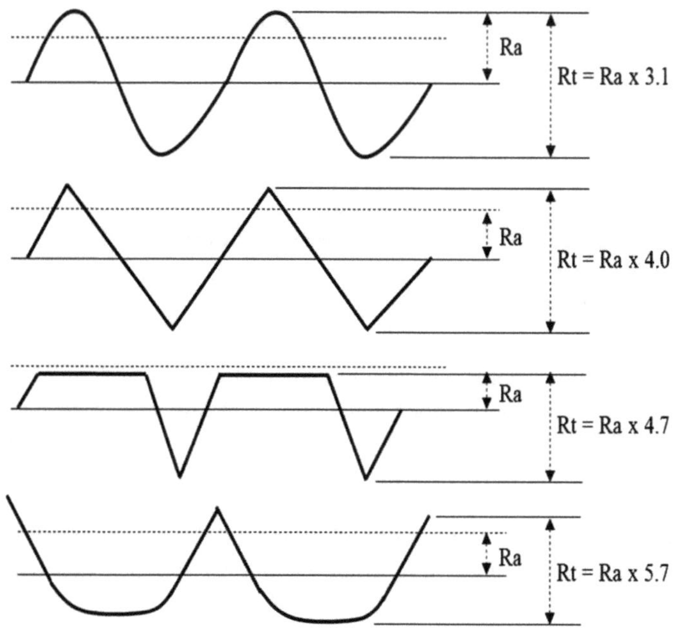

Fig. 4.1 A schematic example of surface profiles that have identical $\boldsymbol{R_a}$ values, but different shapes and values of $\boldsymbol{R_t}$ roughness parameter (Stout and Davis 1984)

Harris and Shaw (1988) also pointed out that it was perhaps more practicable to measure the main parameters that governed slip resistance rather than to measure it directly. Their conclusion implied that systematic measurements of both surface roughness and COFs were necessary to provide reliable results for the slip resistance assessments under dry and contaminated surface conditions.

Jones et al. (1995) described that an absolute COF value could only be useful if the surface roughness of floors and shoe heels/soles were quantified by measuring their topographic features. They demonstrated that micro-roughness determined grips at the shoe-floor sliding interface.

Jones et al. (1995) also suggested developing better methods for measuring surface roughness because the currently available electronic instruments did not take full account of surface roughness profiles. As a result, COFs were influenced by several factors such as test methods (Strandberg 1985), sole patterns (Perkins 1985; Irvine 1970), types of floor surfaces and lubricants, wear of shoe soles, and surface roughness (Manning et al. 1983, 1990; Manning and Jones 1994; Grönqvist 1995; Kim and Smith 2000; Kim et al. 2001, 2013; Kim 2015c, 2016a, b).

Friction mechanisms between the shoe heel and floor surface are much more complex than the classical laws of fiction. As specified by several studies in the literature (Perkins and Wilson 1983; Strandberg 1983; Andres and Chaffin 1985; Tisserand 1985; Kim 1996a, b, c, d, 2004a, b, 2015c; Kim and Smith 1998b, 2001a, b, c; Chang 2000a, b; Kim et al. 2001, 2013; Kim and Nagata 2008a, b), the occurrence and severity of a slip depend not just on the friction when the slip starts, but on how the friction varies as a slip progresses.

As also pointed out by Kim and Smith (1998a) and Kim (2004a, b), observations on the relative surface couplings between two surfaces of the shoe ad floor during dynamic friction processes would provide rather reliable assessments on the slip resistance properties than a simple friction measurement.

4.3 Studies on Surface Roughness Measurements

4.3.1 Background

The recent literature reports that there has been an increasing focus on the surface roughness measurements of shoes and floors with slip resistance measurements (Grönqvist 1995; Rowland et al. 1996; Kim 1996a, b, c, d, 2004a, b, 2015a, b, c, 2016a, b; Chang 1998, 1999, 2000a, b; Manning et al. 1998; Kim and Smith 1998b, 2001a, b, c; Kim et al. 2001, 2013; Kim and Nagata 2008a, b). These studies suggested that the surface features of both shoes and floors substantially influenced the friction mechanisms under a range of walking environments encountered.

Rowland et al. (1996) and Manning et al. (1998) reported the relationship between slip resistance properties and surface roughness of footwear. Rowland et al. (1996) exclusively examined the surface topographies of shoe soles using a scanning electron microscope. They compared microscopic images of the worn shoe surfaces with surface roughness parameters. Although the microscopic approaches were limited, this seemed to be one of the first studies to examine micro-structures of the heel surfaces from a three-dimensional approach.

Chang (1998, 1999) reported a relationship between slip resistance properties and surface roughness of unglazed quarry tiles. Both studies simply claimed that surface roughness of the footwear was less critical than the floor one because the footwear materials were usually not as hard as the floor ones. However, this claim underestimated following two major aspects of tribo-physical properties between the shoe and floor:

(1) Geometric interactions between the surfaces of shoes and floors.
(2) Sliding friction induced wear developments on the shoe and floor surfaces.

The above pointed two tribological events seem to be equally important features observed at the sliding friction mechanism between the shoe heel and floor surface with the friction measurement. Although Chang's two studies showed the effects of surface roughness of shoes and floors on slip resistance results, the following important issues of tribo-physical characteristics were not explored at all:

(1) the slip resistance results were simply compared with surface roughness parameters. As a result, a single aspect of shoe or floor was focused on rather than considering them concurrently.
(2) Simulated polishing methods using an orbital sanding machine and hand polishing with sandpapers did not correctly reproduce natural wear events to generate worn surface conditions. Selection of a pin-on-disk tester for measuring the friction did not adequately simulate the human walking as well.
(3) Wear observation of both bodies in relation to surface topographies was not fully investigated.

Other studies (Kim 1996a, b, c, d; Kim and Smith 1998a, b, 1999, 2001c) also showed the relationships between slip resistance properties and surface roughness parameters. Those studies clearly stated that slip resistance performance between the shoe heels and floor surfaces was mainly influenced by the manner in which the surface topographies of sliding surfaces were changed. Those studies concluded that

(1) Floor surfaces were modified by transfers of polymeric materials from the heel surfaces to the floor ones.
(2) Surface roughness parameters underwent large variations during the initial state of sliding friction, but they stabilised as soon as the steady state of frictional behaviours was achieved.
(3) The mean depth of asperities and average asperities showed the largest alterations and asperity slope angles varied linearly with the $\boldsymbol{R_a}$ roughness parameter.

New theoretical concepts and models on material transfers (or embedding) from the heel surface to floor one were introduced by the recent studies (Kim and Smith 1998a, b, 1999, 2001b, c; Kim et al. 2001, 2013; Kim 1996a, b, c, 2004a, b, 2015a, b, c, d; Kim and Nagata 2008a, b). Those studies investigated the relationships amongst slip resistance properties, surface changes, and wear developments between the shoes and floors. They identified that repeated rubbings during dynamic friction measurements largely changed surface topographies of both shoes and floors.

Alterations on the shoe surfaces were severe than expected and occurred from a very early stage of rubbings. Variations of the surface topographies of both bodies accordingly affected sliding friction mechanism and eventually slip resistance performance. These studies also found out that there were non-linear results between the slip resistance performance and surface roughness under dry and mildly polluted environments such water wet and soapy conditions (Kim et al. 2013; Kim 2015a, b, c; 2016a, b).

Kim and Smith (1998a, 2001c) presented a new concept to analyse slip resistance properties. They investigated the changes of surface roughness within the shoe-floor contact areas, as well as relative surface couplings of the two bodies during repetitive dynamic friction measurements.

To quantify geometric couplings between the shoe and floor surfaces, a concept of comparative geometry mating and index were developed and formulated to characterise surface conformities at the contact areas. In order to verify the concept and index, experimental results showed that the relative surface pairing was well correlated with the results of slip resistance performance.

4.3.2 Measuring Devices for Surface Roughness

Conventional methods for surface roughness measurements have commonly involved with surface profilometers. Any stylus-type profilometer, however, has severe limitations in terms of detectability of detailed surface features and difficulties to generating three-dimensional surface data. To overcome such limitations, alternative instruments and apparatus have been explored to measure topographic features of the shoe and floor surfaces.

Table 4.1 summarises a list of the advanced techniques and technologies for surface analyses which are commonly adopted to measure engineering surface features in the literature. It contains brief information on each instrument with major drawbacks and limitations to apply for the analyses of shoe and floor surfaces.

Table 4.1 A list of instruments for the measurements of surface texture

Type	Brief description	Major drawback
Optical microscope	This is the best-known method for observing surfaces. It reveals fine topographic features with sufficient magnifications but suffers from a number of drawbacks. The use of visible light restricts resolutions of the instrument so it is unable to discriminate features that are smaller than 0.25 μm (Culling 1974; Wilson 1985). This device also lacks the focus of depth so that it tends to emphasise the spacing of departures rather than their actual height.	It is difficult to obtain quantitative values of the size for surface features
Electron microscope	This microscope uses a beam of mono-energetic electrons to produce an image on a fluorescent screen and gives a finer resolution than the optical one, about 10 Å (1 nm), which is about 250 times finer than the best optical one (Hren et al. 1979). Like the optical microscope, this one only examines a very small part of the surface and may mispresent its general characteristics. This deficiency can be overcome in a scanning electron microscope which produces composite images of a larger surface area by allowing a beam to scan in a controlled sequence (Hillmann et al. 1984)	It is difficult to obtain quantitative values of the size for surface features
Interferometry	This instrument gives a good representation of the surface texture and is self-calibrating because it is known that the vertical distance between adjacent fringes represents a half of the wavelength of the light used, about 0.25 μm	This device only allows to viewing a very small and unrepresentative sample of the surface
X-ray diffraction	This instrument is of interest for the determination of crystal and grain orientation and surface deformation. The necessary electrons can be readily adapted to most commercial scanning electron microscopes	This instrument also can not provide quantitative values of the size of surface characteristics

(continued)

Table 4.1 (continued)

Type	Brief description	Major drawback
Electron Micro-probe analyzer	This instrument scans a surface with a 1 μm cross-section electron beam and records the chemical constituents of the surface (Rivière 1983). When a surface is in contact with another surface, this analyser enables to determine the degree of transfer of wear materials from one surface to the other	This instrument can not measure quantitative values of surface characteristics
Laser scanning confocal microscope	This microscope has the potential to revolutionise surface analysis. This instrument is similar to conventional confocal microscope, but it produces better lateral resolutions than the conventional optical one (Wilson and Sheppard, 1984; Wilke 1985; Wilson 1985; van der Voort et al. 1987; White and Amos 1987; Wilson and Carlini 1988; Hook and Odeyale 1989). The distinct advantage of this new microscope is its ability to generate rapidly non-destructive optical sections in thick opaque specimens. Another advantage is that any measuring sample does not need special preparations	It is not easy to use and requires more practice to become skilful in the operation than other instruments

4.4 Understanding of the Shoe-Floor Sliding Friction Interface

Tribology is the science of friction, wear, and lubrication. This field of science was defined in 1967 by a committee of the Organization for Economic Cooperation and Development (OECD). The term “tribology” is derived from the Greek word ‘tribes’, which means rubbing or sliding (Stachowiak and Batchelor 2005; Stachowiak et al. 2005). Its application is widely seen in bearing designs, manufacturing processes, and biomedical applications (e.g., artificial joints). This book applies tribological concepts to understand frictional and wear behaviours of the shoes and floors and their interactive effects on slip resistance problems at the shoe-floor sliding interface. Following Sects. 4.4 and 4.5 discuss basic tribological characteristics at the contact sliding interface between the shoe heel and floor surface during friction measurements.

4.4.1 Significance of Friction Process

Friction is not considered as an intrinsic material property, but rather a characteristic of the pertinent tribological system that includes contacting surfaces, interfacial mediums, and environments encountered. Friction is the prominent reason of energy dissipation whilst wear is the major cause of material depletion (Krim 2012). In contrast, the presence of friction at the shoe-floor interface up to some extent can be extremely useful to sustain walking without slipping and/or falling. Otherwise, if there is little or no friction available between the shoe sole/heel and the floor surface, then fall incidents from slips are likely to take place. Hence, increasing friction up to some levels seems to be desirable in preventing slip and fall incidents.

A common way of assessing slipperiness is to measure traction or slip resistance properties with a format of the coefficient of friction (COF) quantity at the sliding interface between the shoe and floor using a tribometer (Beschorner et al. 2009). The minimum essential friction at the shoe-floor interface to sustain human locomotion is known as required COF (RCOF). Slips usually occur when the COF between the shoe and floor surface (or available COF: ACOF) is less than the required friction (RCOF) needed to continue walking (Hanson et al. 1999; Li et al. 2009).

Fiction and wear behaviours of the shoes and floors and associated tribo-physical characteristics at the sliding interface between the shoe heel and floor surface contain very complex mechanisms and multi-factorial events. Therefore, extensive studies are required to fully understand their tribological activities. Once the underlying friction and wear behaviours and mechanisms are identified and understood, this information can be applied to improve the validity and reliability of slip resistance measurements between the shoes and floors. Accordingly, shoe and floor manufacturers would get a benefit to optimise their designs for shoe treads and floor surface finishes in order to prevent slip and fall incidents.

4.4.2 Measuring Slipperiness

A slip meter typically measures COFs between the shoe and floor surfaces. The measured COF quantity or index indicates a quantity that can be supported without falls at the shoe-floor sliding interface and referred as an available COF (ACOF). The COFs are largely influenced by a number of factors such as shoes, floors, and environmental circumstances such as dry or polluted condition.

The factors affecting the COFs can be broadly categorised as the person- and environment-specific components. For example, the choice of footwear can be categorised as a person-specific component whereas flooring and the presence of a contaminant are considered as an environment-specific component. The environmental factor like the surface roughness of the floor can significantly affect the COF.

Another example for the environment-specific component is the change of viscosity in the fluid contaminant. The change of viscosity may increase or decrease the COF results depending upon the lubrication regime at the shoe-floor sliding interface. Because the environment-specific component can be optimised by technical improvements, this book is mainly focused on the environmental factors.

Furthermore, a sliding speed needs to be taken into account for COF measurements. Some studies in the literature have set a threshold of maximum slipping velocity between 0.5 and 1.0 m/s for an actual slip (Strandberg and Lanshammar 1981; Brady et al. 2000; Cham and Redfern 2002; Beschorner 2008). This book also adopted the sliding speed in the threshold of around 0.4–0.5 m/s during the dynamic sliding friction tests.

4.4.3 Measuring Devices for Slip Resistance

As discussed in Chap. 2, there is a numerous range of apparatuses and devices to quantify frictional behaviours between the shoe and floor. Laboratory devices are sometimes capable of controlling sliding velocities and contact strike angles between the shoe heel and floor surface that are considered most pertinent to human walking such as the Programmable Slip Resistance Tester (PSRT) and Portable Slip Simulator (PSS).

Moreover, most of them can apply for a large magnitude of forces by the attached shoe specimen during slip resistance measurements (Beschorner 2008). According to Aschan et al. (2005), the Portable Slip Simulator enables biomechanically and tribo-physically valid measurements of the slipperiness against outdoor walking surfaces and workplace floors.

Two of the most commonly adopted slip testing devices are the Brungraber Mark II slip meter and the English XL tribometer. They are shown to yield different results for the same test. They are also known as a portable inclinable articulated strut slip tester (PIAST) and a variable incidence tribometer (VIT), respectively. The Brungraber Mark II is a gravity driven device whereas the English XL is a pneumatically driven one.

However, both devices measure frictional behaviours, based on the slipping or sticking at the moment when a shoe sample strikes a floor surface. These slip meters allow for a collision, or in other words, an impact between the shoe material and the floor surface at variable angles and COFs are quantified by the angles that the shoe material transits from sticking to the floor surface to slipping out (Chang and Matz 2001; Chang et al. 2003; Beschorner 2008).

Grönqvist et al. evaluated portable slip testers and concluded that the Brungraber Mark II possessed a highly variable normal force and an uncontrolled sliding velocity during the slip tests (Grönqvist et al. 1999). Moreover, measuring characteristics can be different amongst the slip meters (Chang et al. 2003). The discrepancies amongst each measurement lead to controversies between the qualities of data recorded from the different slip meters (Beschorner 2008).

Another issue to consider is the test protocols for using the slip meters. Even though there are protocols provided by the ASTM International, the operations of the slip meters are somehow subjective. The measurements may vary from one operator to another. Therefore, it can be considered that measuring devices for the pedestrian fall safety assessment still require significant improvements to be universally accepted.

4.5 Basic Tribology for the Shoe-Floor Sliding Friction Mechanism

4.5.1 Pedestrian Slip Resistance Requirements

Slip resistance is the ability of a surface to substantially reduce or prevent the risk of falling (C&CAAO 2003). It is generally termed to describe those textured surfaces of shoes and floors that perform well in preventing falls under dry and lubricated environments. Research carried out by the Building Research Station (BRS) in the UK shows that for normal ambulatory activities the pedestrians require a range of DFCs between the shoes and the floors (The Tile Association 2016). They found that people required different levels of slip resistance at different times within that range. Based on the BRS work and analysis of slip and fall incidents, it is commonly accepted that a COF quantity of 0.4 is required for safe walking without falling. Only in certain well-defined circumstances should this be reduced.

However, slip resistance measurements for the pedestrian fall assessment are a complicated issue where the likelihood of a fall is a function of the multi-faceted factors such as walkways, environments, walkers' physical conditions, footwear, etc. There is an expectation that the footwear and walkways should provide sufficient slip resistance or traction functions during their service periods. Therefore, in-depth knowledge is required to fully understand the complex nature of frictional behaviours between the shoe heel and floor surface. The tribological approach may provide an unbiased alternative to explore the surface features of shoes and floors and their interactive effects on slip resistance performance.

4.5.2 Shoe-Floor Friction and COF Measurements

A COF is developed by interactions between two contact-sliding surfaces. Its nature involves extremely complex interactions between the two bodies and no means by which the COF can be accurately predicted from a general knowledge of the two surfaces concerned.

The COF is also critically dependent upon the presence of a lubricant or contaminant faced. One of the most common types of contaminants encountered is water in our walking environments. Since most but not all the fall accidents occur under the wet condition, slip resistance tests of the floor surface is normally conducted under this environmental situation. If other types of contaminants such as oils and fats are anticipated, for example in the kitchen environment, it may be necessary to use these lubricants in slip tests.

Dry slipping or sliding can also occur in certain situations so it is important to check on this aspect as well. However, it should be stated that the 'classical laws' of friction are based on experimental studies and apply to dry, smooth and rough solid homogeneous surfaces in contact. It has been acknowledged that the laws for dry friction do not fully explain the reason for how traction or grip is developed between the footwear and floor surfaces. It is already well known that the laws that explain dry friction change when the contacting surfaces are lubricated (Jenson and Chenoweth 1990; Meriam and Kraige 1992). Therefore, the friction mechanism would be different to the mechanism explaining dry friction when one of the contacting surfaces is not smooth or solid but is granular and/or is wet, or when the footwear has cleats or studs (Barry and Milburn 2013).

4.5.3 Function of Shoes on Slip Resistance

Most but by no means, all the shoe sole and/or heel materials give good levels of friction against most floor surfaces under clean and dry environments. Some shoe types such as Nylon, hard Polyurethane, and PVC soles/heels show notable exceptions in their slip resistance performance. However, the worst acceptable shoe material is leather. Under the wet condition or where any contamination is involved, absorbent materials such as leather soles and/or heels behave differently in their slip resistance functions as compared with non-absorbent ones such as rubbers.

In the latter case and where the floor surface is relatively smooth, the material properties of rubbers or plastics are far less important than under the dry environment. As a result, the slip resistance functions are largely depended on the roughness of the rubber/plastic surface. Most rubbers tend to wear smoothly, but some special formulations known as microcellular, rubbers, and plastics retain their roughness and provide significantly better slip resistance performance under polluted environments.

4.5.4 Function of Floors on Slip Resistance

It is important to choose and specify the best floors and flooring materials that are not only safe but also holding up and durable over time periods. There are a wide variety of floor surfaces, floor coverings and flooring materials available from the market, most of which give satisfactory slip resistance under clean and dry environments. Under the dry surface condition, however, smooth and shiny floors can give very higher slip resistance against a soft rubber shoe than rough ones. On the other hand, under wet or contaminated environments, the shiny and smooth floors become far less slip resistant compared to the rougher ones, which may only lose a small proportion of its slip resistance performance. A very smooth and shiny floor surface can normally be regarded as dangerous against the wet or lubricated conditions although there are few exceptions.

However, one of the major problems with floor surfaces is to predict their wear and tear developments and effects on slip resistance properties under a range of pedestrian walking environments (Kim et al. 2013; Kim 2015a). With continued ambulation, however, surface finishes of the floors and floor coverings can be changed by a number of causes such as ageing, corrosion, soiling, and maintenance (Leclercq and Saulnier 2002). As a result, progressive wear advances of the floor surfaces are unavoidable and can substantially affect slip-resistance properties (Kim and Nagata 2008b; Kim et al. 2013; Kim 2015c).

This seems to become particularly problematic when slip tests are conducted against new floors and walkways. Because topographic structures of the new underfoot surfaces and walkways may significantly different as compared with the ones from the real world of walking environments through their lifetime services. Despite the importance of flooring wear issue, its fundamental perception and effects on slip-resistance properties have not been systematically discussed in the literature. Above all, unfortunately, it is scarce to find any definitive concepts and/or methodical studies on wear behaviours of pedestrian floors and their influences on slip-resistance functioning.

Another issue to drawing an attention is that whilst increasing the slip resistance properties of floor surfaces would be desirable as a general rule, a very high traction or slip resistance may impede safe and comfortable ambulation (Chaffin et al. 1992; Kim 2015a, b, c, d). Maintaining or increasing the floor surface roughness also requires high processing costs. Even though numerous experimental and analytical studies on the prevention of slip and fall incidents are found in the literature, no theoretical concept and/or model is developed to predict the effect of floor surface finishes on slip resistance performance.

In particular, it is hard to find any definitive study and/or design information for functional ranges of floor surface finishes required for optimal slip resistance performance (Kim et al. 2013; Kim 2015a, d). There are also no internationally accepted guidelines and design data on functional levels of floor surface coarseness for the effective control of slip resistance performance. Therefore, it is necessary to develop a method which can provide practical design information for the floor surface roughness against a range of walking environments.

4.5.5 Factors Affecting Film Formations

The surface roughness of floors and floor coverings is one of the main factors, which prevent liquid film formations between the shoe heel and floor surface.

Hence, surface finishes of the floors and floor coverings should provide optimal levels of roughness scales. But, one of the main difficulties is that most roughness meters cannot differentiate between the floors with rolling hill types of surface topographies and mountainous types of ones. One can get the same readings of surface roughness from two such floors, but the rolling hill type may have a slippery surface whilst the sharp-peaked mountainous type seems to be slip resistant one.

Although surface roughness readings are useful to evaluating floor surface texture, measurements of surface roughness seem to be a small part of the answers to assessing floor surface features. However, it becomes clear from the above discussions that cleanability of the floor surfaces can vary dramatically depending on the nature of surface roughness. The viscosity of liquid contaminants is also an important matter to consider. In general, the more viscous the liquid the thicker would be the lubricating films and the rougher the surface needs to give better slip resistance.

Under the clean and dry environments, the main contaminant sources seem to dust particles and/or dirt. As yet, there are no "standard" dust used for slip resistance testing, hence one must use dust collected at the test site although this is clearly not an ideal practice (The Tile Association 2016). Dust can cause significant reductions in slip resistance properties, especially if they contain hard particles and the floor surface is a hard material. In such circumstances, dust particles seem to roll between the floor and the heel surfaces. Therefore, those factors concerning different types of film formations on the floor surface require in-depth investigations to identify their involved tribo-physical characteristics and effects on slip resistance performance.

4.6 Slip Resistance Measurement and Reaction

Slip resistance tests should be accurate and interpretation of measurement results should be comparable with minimum recommendations to justify decisions that involve remedial actions. They also should provide peace of mind knowing that the floor surfaces to comply with the best-practiced recommendations for the fall safety assessments. Corrective actions addressing the floor surfaces that do not meet references need to take into account numerous factors such as tested shoes, surrounding environments, chemical resistance/attack on the floor surfaces and surface materials. Therefore, it is recommended that compelling advice from an independent floor consultant needs to be sought.

However, a floor surface that has a significantly higher slip resistance property than the recommendations may present other safety and hygiene issues. As discussed above, floor surfaces with a very high level of traction or slip-resistance property may obstruct secure ambulation and cause potential injuries from trips and resulting abrasions (Chaffin et al. 1992; Kim et al. 2013). Another issue to consider is that a floor surface with an overly high slip resistance property may be so rough that it can introduce cleaning challenges. This means that floor surfaces with very rough traction properties can also trap contaminants leading to concerns over hygiene matters.

There is little advantage in exceeding the recommendations for slip resistance if cleaning and maintenance of the floor surface become either impractical or unviable. For example, swimming pools can unintentionally introduce bacteria, fungi, viruses, and other pathogens onto the floor surface which may create health problems that can be transferred to pool users (Langdon 2011). Selection of an appropriate cleaning regime is of utmost importance for the floor surfaces of swimming pools.

The task of cleaning may appear to be simple but seems to deal with complex scientific and technical challenges. Without full considerations for the cleaning methods and materials, the floor surface can suffer irreparable damage or the contaminants present may not be removed, thereby creating potential hygiene issues. It follows therefore that the detailed studies comparing cleaning regimens and processes in the setting of slip resistance performance are immediately required to address these issues.

4.7 Conflict over Slip Resistance, Hygiene, and Maintenance

Poor designs and inappropriate maintenance/cleaning methods for polluted floors and floor coverings would be responsible for property owners and managers with consequences that can have far-reaching implications for safety and costs (Langdon 2011). For example, wet areas like pool surrounds and adjacent facilities such as changing and shower rooms present potential slip and fall hazards that are often an afterthought of design or occur with increasing regularity over time periods as the surface becomes more slippery. Hence, the first challenge is to identify slip resistance properties between the shoes and floor surfaces under such specific environments. Slip resistance needs to be routinely measured and monitored and regular maintenance should be effective and efficient.

The second challenge is to realise the consequences of installing an overly aggressive surface that may exceed recommendations for slip resistance, but introduce other safety and health issues resulting from an overly rough floor surface. The risk of infection due to poor hygiene and/or cleaning also becomes a factor that needs to be properly addressed. Because floors and floor coverings can serve as an important reservoir for infectious organisms, infection control is a matter of great concern and a major challenge to the healthcare environments such as hospitals. Currently, available cleaning methods and techniques for the removal of contamination are inadequate (Abreu et al. 2013). Therefore, developments of new disinfectant techniques and policies are urgently required for the prevention of flooring-attributable infectious disease.

4.8 Chapter Summary

An adequate level of traction or slip resistance at the shoe-floor sliding interface is required for unperturbed ambulation. Without the presence of friction, safe ambulation just could not occur. The classical model of friction simply defines slip resistance properties with simple force variables and is limited in its ability to explain the mechanics presented at the shoe-floor sliding interface.

In the context of human ambulation, however, friction is a complex phenomenon and contains multi-factorial mechanisms. Over time periods, the classical model of friction has developed into a paradigm which accounts for both human- and environmental-specific components. However, surface topographies of both shoe and floor can be largely modified by friction-induced wear developments. As a result, this would significantly affect slip resistance properties. Therefore, surface finishes of the shoe and floor should be monitored routinely to maintain and provide the best slip resistance performance against specific walking/working environments.

Therefore, surface finishes of the shoe and floor should be monitored routinely to maintain and provide the best slip resistance performance against specific walking/working environments. This information may provide more reliable results to manage pedestrian fall safety than measuring slip resistance alone. Such approaches highlight the need for developing enhanced concepts and methods for reliably characterizing slip resistance properties. This should be based on thorough understanding of the complex nature of frictional behaviours between the shoe and floor surface, their related tribo-physical characteristics, and their interactive impacts on slip resistance performance.

References

Abreu, A. C., Tavares, R. R., Borges, A., Mergulhão, F., & Simões, M. (2013). Current and emergent strategies for disinfection of hospital environments. *Journal of Antimicrobial Chemotherapy, 68*(12), 2718–2732.

Andres, R. O., & Chaffin, D. B. (1985). Ergonomic analysis of slip-resistance measurement devices. *Ergonomics, 28*(7), 1065–1080.

Aschan, C., Hirvonen, M., Mannelin, T., & Rajamäki, E. (2005). Development and validation of a novel portable slip simulator. *Applied Ergonomics, 36*(5), 585–593.

Barry, B., & Milburn, P. (2013). Tribology, friction and traction: understanding shoe-surface interaction. *Footwear Science, 5*(3), 137–145.

Beschorner, K. E. (2008). Development of computational model for shoe-floor-contaminant friction. PhD dissertation, Swanson School of Engineering, University of Pittsburgh, USA.

Beschorner, K. E., Lovell, M., Higgs, C. F., III, & Redfern, M. S. (2009). Modeling mixed-lubrication of a shoe-floor interface applied to a pin-on-disk apparatus. *Tribology Transactions, 52*(4), 560–568.

Bhushan, B., Wyant, J. C., & Meiling, J. (1988). A new three-dimensional non-contact digital optical profiler. *Wear, 122*(3), 301–312.

Brady, R. A., Pavol, M. J., Owings, T. M., & Grabiner, M. D. (2000). Foot displacement but not velocity predicts the outcome of a slip induced in young subjects whilst walking. *Journal of Biomechanics, 33*(7), 803–808.

Braun, R., & Roemer, D. (1974). Influence of waxes on the static and dynamic friction. Soap/Cosmetics/Chemical Specialties, December, pp. 60–72.

Bring, C. (1982). *Testing of slipperiness. Document D5.* Stockholm, Sweden: The National Swedish Council for Building Research.

Cement & Concrete Association of Australia Offices (C&CAAO). (2003). Slip resistance of residential concrete paving surfaces. July 1–8. Available from: http://59.167.233.142/publications/pdf/profSlip.pdf.

Chaffin, D. B., Woldstad, J. C., & Trujillo, A. (1992). Floor/shoe slip resistance measurement. *J Am Indust Hyg Assoc, 53*(5), 283–289.

Cham, R., & Redfern, M. S. (2002). Heel contact dynamics during slip events on level and inclined surfaces. *Safety Science, 40*(7–8), 559–576.

Chang, W. R. (1998). The effects of surface roughness on dynamic friction between neolite and quarry. *Safety Science, 29*(2), 89–105.

Chang, W. R. (1999). The effects of surface roughness on the measurement of slip resistance. *International Journal of Industrial Ergonomics, 24*(3), 299–313.

Chang, W. R. (2000a). The effect of filtering processes on surface roughness parameters and their correlation with the measured friction, Part II: Porcelain tiles. *Safety Science, 36*(1), 35–47.

Chang, W. R. (2000b). The effects of surface roughness and contaminants on the dynamic friction between porcelain tile and vulcanized rubber. in *Proceeding of the 14th Triennial Congress of the International Ergonomics Association, IEA 2000/HFES 2000*, August, San Diego, California, USA, No. 4, pp. 494–497.

Chang, W. R., & Matz, S. (2000). The effect of filtering processes on surface roughness parameters and their correlation with the measured friction, Part I: Quarry tiles. *Safety Science, 36*(1), 19–33.

Chang, W. R., & Matz, S. (2001). The slip resistance of common footwear materials measured with two slipmeters. *Applied Ergonomics, 32*(6), 549–558.

Chang, W. R., Cotnam, J. P., & Matz, S. (2003). Field evaluation of two commonly used slipmeters. *Applied Ergonomics, 34*(1), 51–60.

Chang, W. R., Grönqvist, R., Leclercq, S., Myung, R., Makkonen, L., Strandberg, L., et al. (2001). The role of friction in the measurement of slipperiness, Part 1: Friction mechanisms and definition of test conditions. *Ergonomics, 44*(13), 1217–1232.

Culling, C. F. A. (1974). *Modern microscopy, elementary theory and practice.* London: Butterworth.

Grönqvist, R. (1995). Mechanisms of friction and assessment of slip resistance of new and used footwear soles on contaminated floors. *Ergonomics, 38*(2), 224–241.

Grönqvist, R., Hirvonen, M., & Tohv, A. (1999). Evaluation of three portable floor slipperiness testers. *International Journal of Industrial Ergonomics, 25*(1), 85–89.

Grönqvist, R., Hirvonen, M., & Skytta, E. (1992). Countermeasures against floor slipperiness in the food industry. in S. Kumar (Ed.), *Advances in Industrial Ergonomics and Safety IV* (pp. 989–995). Taylor and Francis.

Grönqvist, R., Roine, J., Järvinen, E., Korhonen, E., & Rahikainen, A. (1990). Slip resistance versus surface roughness of deck and other surfaces in ships. *Journal of Occupational Accidents, 13,* 291–302.

Hanson, J. P., Redfern, M. S., & Mazumdar, M. (1999). Predicting slips and falls considering required and available friction. *Ergonomics, 42*(12), 1619–1633.

Harris, G. W., & Shaw, S. R. (1988). Slip resistance of floors: Users' opinions, Tortus instrument readings and roughness measurement. *Journal of Occupational Accidents, 9*(4), 287–298.

Hillmann, W., Kranz, O., & Eckolt, K. (1984). Reliability of roughness measurements using contact stylus instruments with particular reference to results of recent research at the physikalisch-technique bundesanstalt. *Wear, 97*(1), 27–43.

Hoang, K., Stevenson, M. G., & Wilgoss, R. A. (1985). Measurement of dynamic friction between shoe soles and walkway surfaces. In *Proceedings of the 22nd Annual Conference of the Ergonomics Society of ANZ*, Toowoomba, December, pp. 265–271.

Hook, G. R., & Odeyale, C. O. (1989). Confocal scanning fluorescence microscopy: A new method for phagocytosis research. *Journal of Leukocyte Biology, 45*(4), 277–282.

Hren, J. J., Goldstein, J. I., & Joy, D. C. (1979). *Introduction to analytical electron microscopy*. New York: Plenum Press.

Hunter, R. N. (1930, August). A method of measuring frictional coefficients of walkway materials. *National Bureau of Standards Journal of Research*, *3*, 329–348.

Irvine, C. H. (1970). Shoe sole slipperiness on structural steel. *Materials Research and Standards, 10*(4), 21–22.

Jenson, A., & Chenoweth, H. H. (1990). *Applied engineering mechanics* (p. 164). Ohio: Glencoe/McGraw-Hill.

Jones, C., Manning, D. P., & Bruce, M. (1995). Detecting and eliminating slippery footwear. *Ergonomics, 38*(2), 242–249.

Jung K., & Reidiger, G. (1982). Recent developments regarding the inspection of non-slip floor coverings. Die Berufsgenossenschaft, 1–7; Health and safety executive translation number 12053.

Kennaway, A. (1970). On the reduction of slip of rubber crutch tips on wet pavements, snow and ice. *Bulletin of Prosthetics Research, Fall*, pp. 130–144.

Kim, I. J. (1996a). Tribological concepts for the investigation of the pedestrian slipping and falling accidents—Part I. In *International Occupational Injury Symposium*, February, Sydney, Australia, p. 101.

Kim, I. J. (1996b). Tribological approach for the analysis of pedestrian slip hazard-II. In *Proceedings of the '96 Spring Conference of Korean Institute of Industrial Engineers*, April, Seoul, Korea, pp. 279–285.

Kim, I. J. (1996c). Microscopic investigation to analyze the slip resistance of shoes. In *Proceedings of the 4th Pan Pacific Conference on Occupational Ergonomics*, November, Taiwan, ROC, pp. 68–73.

Kim, I. J. (1996d). Microscopic observation of shoe heels for pedestrian slip hazard investigation. In *Proceedings of the 1st Annual International Conference on Industrial Application and Practice*, December, Texas, U.S.A., pp. 243–250.

Kim, I. J. (2004a). Development of a new analyzing model for quantifying pedestrian slip resistance characteristics: Part I—Basic concepts and theories. *Industrial Journal of Industrial Ergonomics, 33*(5), 395–401.

Kim, I. J. (2004b). Development of a new analyzing model for quantifying pedestrian slip resistance characteristics: Part II—Experiments and validations. *Industrial Journal of Industrial Ergonomics, 33*(5), 403–414.

Kim, I. J. (2006a). The current hiatus in fall safety measures. In W. Karwowski (Ed.), *International encyclopedia of ergonomics and human factors-2005* (pp. 2572–2576). USA: Taylor & Francis Group, LLC.

Kim, I. J. (2006b). A new paradigm for characterizing slip resistance properties. In W. Karwowski (Ed.), *International Encyclopedia of Ergonomics and Human Factors-2005* (pp. 2735–2740). USA: Taylor & Francis Group, LLC.

Kim, I. J. (2015a). Practical design search for optimal floor surface finishes to prevent fall incidents (Chapter 5). In Evans, B. (Ed.), *Accidental falls: Risk factors, prevention strategies and long-term outcomes* (pp. 80–103). Hauppauge, NY 11788, USA: Nova Science Publishers, Inc.

Kim, I. J. (2015b). Slip-resistance measurements for assessing pedestrian falls: Facts and fallacies (Chapter 6). In Evans, B. (Ed.), *Accidental falls: Risk factors, prevention strategies and long-term outcomes* (pp. 105–125). Hauppauge, NY 11788, USA: Nova Science Publishers, Inc.

Kim, I. J. (2015c). Wear observation of shoe surfaces: Application for slip and fall safety assessments. *Tribology Transactions, 58*(3), 407–417.

Kim, I. J. (2015d). Research challenges on slip-resistance measurements for assessing pedestrian fall incidents. *Journal of Ergonomics, 5*(3). doi:10.4172/2165-7556.1000e142

Kim, I. J. (2016a). A study on wear development of floor surfaces: Impact on pedestrian walkway slip-resistance performance. *Tribology International, 95,* 316–323.

Kim, I. J. (2016b). Identifying shoe wear mechanisms and associated tribological characteristics: The importance for slip resistance evaluation. *Wear, 360–361,* 77–86.

Kim, I. J., & Nagata, H. (2008a). Nature of the shoe wear: Its uniqueness, complexity and effects on slip resistance properties. *Contemporary ergonomics* (Vol. 15, pp. 728–734). UK: Taylor & Francis.

Kim, I. J., & Nagata, H. (2008b). Research on slip resistance measurements—A new challenge. *Industry Health, 46*(1), 66–76.

Kim, I. J., & Smith, R. (1998a). A study of the comparative geometry mating between the surfaces of the shoe and floor in pedestrian slip resistance measurements. In *The 5th Pan-Pacific Conference on Occupational Ergonomics*, July, Kitakyushu, Japan, pp. 34–37.

Kim, I. J., & Smith, R. (1998b). Tribological characterization of the frictional force component in pedestrian slip resistance measurements. In *Third World Congress of Biomechanics (WCB '98)*, August, Hokkaido, Japan.

Kim, I. J., & Smith R. (1999). The relationship between wear, surface topography characteristics and coefficient of friction as a means of assessing the slip hazards. In *2nd Asia-Pacific Conference on Industrial Engineering and Management Systems (APIEMS'99)*, October, Ashikaga, Japan, pp. 155–161.

Kim, I. J., & Smith, R. (2000). Observation of the floor surface topography changes in pedestrian slip resistance measurements. *Industrial Journal of Industrial Ergonomics, 26*(6), 581–601.

Kim, I. J., & Smith, R. (2001a, June). A critical analysis on the friction measuring concept for slip resistance evaluation. In *ASTM Symposium on the Metrology of Pedestrian Locomotion and Slip Resistance* (pp. 1–14). West Conshodocken, Pennsylvania, USA: ASTM Headquarters.

Kim, I. J., & Smith, R. (2001b, August). A study for characterising topography changes of shoe surfaces in the early stage of slip resistance measurements—Bearing area curve. In *6th Pan-Pacific Conference on Occupational Ergonomics, Beijing, P. R. China*, pp. 299–303.

Kim, I. J., & Smith, R. (2001c, August). Three-dimensional analysis of floor surface wear during slip resistance measurements. In *6th Pan-Pacific Conference on Occupational Ergonomics*, Beijing, P. R. China, pp. 304–308.

Kim, I. J., & Smith, R. (2003). A critical analysis of the relationship between shoe-floor wear and pedestrian/walkway slip resistance. In M. I. Marpet & M. A. Sapienza (Eds.), *Metrology of pedestrian locomotion and slip resistance (American Society of Testing and Materials, Special technical publication 1424)* (pp. 33–48). Philadelphia, USA: ASTM International.

Kim, I. J., Hsiao, H., & Simeonov, P. (2013). Functional levels of floor surface roughness for the prevention of slips and falls: Clean-and-dry and soapsuds-covered wet surfaces. *Applied Ergonomics, 44*(1), 58–64.

Kim, I. J., Smith, R., & Nagata, H. (2001). Microscopic observations of the progressive wear on the shoe surfaces which affect the slip resistance characteristics. *Industrial Journal of Industrial Ergonomics, 28*(1), 17–29.

Kime, G. A. (1991). Slip resistance and the UK slip resistance research group. *Safety Science, 14*(3–4), 223–229.

Krim, J. (2012). Friction and energy dissipation mechanisms in adsorbed molecules and molecularly thin films. *Advances in Physics, 61*(3), 155–323.

Langdon, S. (2011). Wet floors. The battle over slip, hygiene and maintenance. *Advanced Technology Testing and Research.* September 23. Available from: http://www.attar.com.au/ckfinder/userfiles/files/MAV_article_Final_web.pd

Leclercq, S., & Saulnier, H. (2002). Floor slip resistance changes in food sector workshops: prevailing role played by "fouling". *Safety Science, 40*(7), 659–673.

Li, K. W., Chang, W. R., & Chang, C. C. (2009). Evaluation of two models of a slip meter. *Safety Science, 47*(10), 1434–1439.

Manning, D. P., & Jones, C. (1994). The superior slip-resistance of footwear soling compound T66/103. *Safety Science, 18*(1), 45–60.

Manning, D. P., Jones, C., & Bruce, M. (1983). Improved slip resistance on oil from surface roughness of footwear. *Rubber Chemistry and Technology, 56*(4), 703–717.

Manning, D. P., Jones, C., & Bruce, M. (1985). Boots for oily surfaces. *Ergonomics, 28,* 1011–1019.

Manning, D. P., Jones, C., & Bruce, M. (1990). Proof of shoe slip-resistance by a walking traction test. *Journal of Occupational Accidents, 12*(4), 255–270.

Manning, D. P., Jones, C., Rowland, F. J., & Roff, M. (1998). The surface roughness of a rubber soling material determines the coefficient of friction on water-lubricated surfaces. *Journal of Safety Research, 29*(4), 275–283.

Meriam, J. L., & Kraige L. G. (1992). *Engineering mechanics volume 1 statics* (p. 346). New York, NY: John Wiley.

Nagata, H. (1993). Fatal and non-fatal falls—A review of earlier articles and their developments. *Safety Science, 16,* 379–390.

Perkins, P. (1985, May). Sole design guidelines developed to improve slip-resistance. *SATRA Bulletin*, pp. 58–59.

Perkins, P. J., & Wilson, M. P. (1983). Slip resistance testing of shoes—New development. *Ergonomics, 26*(1), 73–82.

Proctor, T. D. (1993). Slipping accidents in Great Britain—An update. *Safety Science, 16*(3–4), 367–377.

Proctor, T. D., & Coleman, V. (1988). Slipping and tripping accidents and falling accidents in Great Britain—Present and future. *Journal of Occupational Accidents, 9*(4), 269–285.

Rivière, J. C. (1983). Practical surface analysis by auger and x-ray photoelectron spectroscopy. In D. Briggs & M. P. Seah (Eds.), *Instrumentation*. Chichester, England: Wiley (Chapter 2).

Rowland, F. J., Jones, C., & Manning, D. P. (1996). Surface roughness of footwear soling materials: Relevance to slip-resistance. *Journal of Testing and Evaluation, 24*(6), 368–376.

Sherrington, I., & Smith, E. H. (1987). Parameters for characterizing the surface topography of engineering components. *Proceedings of the Institution of Mechanical Engineers, Part C: Journal of Mechanical Engineering Science, 201*(4), 297–306.

Sigler, A., Geib, M. N., & Boone, T. H. (1948, May). Measurement of the slipperiness of walkway surfaces research paper RP 1879. *Journal of Research of the National Bureau of Standards, 40,* 339–346.

Stachowiak, G. W., & Batchelor, A. W. (2005). *Engineering tribology*. Burlington, MA, USA: Butterworth-Heinemann.

Stachowiak, G. P., Podsiadlo, P., & Stachowiak, G. W. (2005). A comparison of texture feature extraction methods for machine condition monitoring and failure analysis. *Tribology Letters, 20*(2), 133–147.

Stevenson, M. G., Hoang, K., Bunterngchit, Y., & Lloyd, D. G. (1989). Measurement of slip resistance of shoes on floor surfaces, Part 1: Methods. *Journal of Occupational Health and Safety, 5*(2), 115–120.

Stout, K. J., & Davis, E. J. (1984). Surface topography of cylinder bores-the relationship between manufacture, characterization and function. *Wear, 95*(2), 111–125.

Strandberg, L. (1983). On accident analysis and slip resistance measurement. *Ergonomics, 26*(1), 11–32.

Strandberg, L. (1985). The effect of conditions underfoot on falling and overexertion accidents. *Ergonomics, 28*(1), 131–147.

Strandberg, L., & Lanshammar, H. (1981). The dynamics of slipping accidents. *Journal of Occupational Accidents, 3*(3), 153–162.

The Tile Association. (2016). *Slip resistance of hard flooring*. Technical Publication. Available from: http://www.tiles.org.uk/wp-content/uploads/2016/01/Slip-Resistance-of-Hard-Flooring-Technical-Publication.pdf

Tisserand, M. (1985). Progress in the prevention of falls caused by slipping. *Ergonomics, 28*(7), 1027–1042.

van der Voort, H. T. M., Brakenhoff, G. T., & Janssen, G. C. A. M. (1987). Determination of the 3-dimensional properties of a confocal scanning laser microscope. *Optik, 78,* 48–53.

White, J. G., & Amos, W. B. (1987). Confocal microscopy comes of age. *Nature, 328*(6126), 183–184.

Wilke, V. (1985). Optical scanning microscopy—The laser scan microscope. *Scanning, 7,* 88–96.

Wilson, M. P., & Perkins, P. J. (1985). Evaluation of a slip-resistance test for shoes. *Ergonomics, 28,* 1081–1085.

Wilson, T. (1985). Scanning optical microscopy. *Scanning, 7,* 79–87.

Wilson, T., & Carlini, A. R. (1988). Three-dimensional imaging in confocal imaging systems with finite sized detectors. *Journal of Microscopy, 149*(1), 51–66.

Wilson, T., & Sheppard, C. (1984). *Theory and practice of scanning optical microscope*. London: Academic Press.

Chapter 5
Friction and Wear Mechanisms

5.1 Introduction

When two solid surfaces are placed together, contact generally occurs only over isolated parts of the nominal contact area (Bhushana 2003). It is through these localised regions of contact that forces are exerted between the two bodies, and it is these forces which are responsible for friction. This means that friction is the resistance to motion that occurs whenever one solid body slides over another (Kim and Smith 2000; Kim 2004a, b, 2006a, 2015a, b, c, d). The resistive force, which acts in a direction directly opposite to the direction of motion, is known as the friction force (or tangential force). The friction force that is required to initiate sliding is known as the static friction force, whilst that required to maintaining sliding is known as the kinetic or dynamic friction force (Moore 1972; James 1980). Kinetic friction is usually lower than static friction.

Because of the great variety of types, forms, and processes of friction and friction induced wear phenomena which determine the behaviours of any tribo-physical entity, it would be necessary to exploring a sub-classification of these important tribological terms involved in the shoe-floor sliding friction mechanism. For this sub-classification, following different criteria are applied:

(1) kinematics or type of motion—contact and sliding;
(2) type of tribology—adhesion, deformation, sloughing, abrasion, etc.; and
(3) type of material—general flooring and footwear materials.

Unfortunately, however, as with the friction process, there is no reliable way of predicting wear development a priori between the two materials to be used under the proposed rubbing conditions.

Wear behaviours can vary over many orders of magnitude and may change catastrophically even though a relatively small change in operating conditions (Czichos 1974). Thus, the wrong choice of materials can have disastrous consequences.

At the most elementary level, wear behaviours can be manifested by a loss of surface material from one or both surfaces when they are subjected to relative motion

I.-J. Kim, *Pedestrian Fall Safety Assessments*,
DOI 10.1007/978-3-319-56242-1_5

(OECD 1968). Sometimes wear progress is clearly visible to the naked eyes; in other cases, it requires the most elaborate measurement techniques for its detection.

Although the wear event mainly occurs at the surface contact that is also the seat of the friction mechanism, the two phenomena (friction and wear) are not simply related one to the other. This means that low friction does not necessarily mean low wear and vice versa.

Wear growth is rarely the result of a single mechanism. A wide variety of conditions causes wear, with many mechanisms such as adhesion, abrasion, corrosion, and fatigue contributing to the damage caused. There are situations where one type changes to another, or where two or more mechanisms operate together.

In any particular instance of the wear event, one may have any of these mechanisms operating either singly or in combination. In the latter case, the situation is even more complicated because of interactions amongst several mechanisms.

As a result of this complicated situation, therefore, wear behaviours should be considered for the analysis of slip safety assessment with friction measurements. However, wear behaviours of the shoe and floor surface are rarely considered in the study of pedestrian fall safety until now.

In this chapter, therefore, the origins of friction forces between the shoe and floor surface are fully discussed and related tribo-physical characteristicssuch as wear behaviours are comprehensively explored as well. However, the discussion is restricted to the friction and wear mechanisms of unlubricated solids in sliding motion as a first step.

5.2 Friction Mechanism

5.2.1 Definition of Friction

Friction is the resistance to motion that occurs whenever one solid body slides over another. The resistive force, which acts in a direction directly opposite of any motion or applied force that is trying to move the object, is known as a friction force. The friction force which is required to initiate sliding is known as astatic friction force, whilst that required to maintaining sliding is known as a kinetic (or dynamic) friction force (Moore 1972; James 1980).

Such broad definition embraces two important classes of relative motion: sliding and rolling. In both ideal situations of rolling and sliding, a tangential force, $\boldsymbol{F}$, is needed to move the upper body over the stationary counterface. A ratio between this frictional force and the normal load, $\boldsymbol{W}$, is known as a coefficient of friction (COF) and is usually denoted by the symbol, $\boldsymbol{\mu}$:

$$\mu = \frac{F}{W} \tag{5.1}$$

The magnitude of the frictional force is conveniently described by the value of the COF, which can vary over a wide range. As with the friction force, it can be defined as a static coefficient of friction (SCOF) and a dynamic coefficient of friction (DCOF). But, unless stated otherwise, the COF quantity is normally taken to designate the DCOF in this book.

5.2.2 The Laws of Friction

A systematic study on frictional behaviours and events goes back to the time of Leonardo da Vinci, but the best known formal statements of this early work are those propounded by Amontons and Coulomb (Moore 1972) and so-called as laws of friction. These laws are entirely empirical and no physical principles are violated in those cases where the laws are not obeyed. The laws of friction may be stated as follows:

(1) The friction force $\boldsymbol{F}$ is proportional to the normal load $\boldsymbol{W}$. The constant of proportionality is termed by the COF, $\boldsymbol{\mu}$. Then,

$$F = \mu \times W \tag{5.2}$$

(2) The friction force is independent of the apparent area of contact.
(3) The friction force is independent of the sliding velocity between the two surfaces.

The above three laws of friction have varying reliabilities, but provide useful summaries of empirical observations except in some important cases. Ever since these laws were formulated in the 17th century, the view has been passed down that for any pair of materials. There are just two COFs, static and dynamic and the static COF is always greater than the dynamic one.

In addition, it is commonly believed that the frictional force cannot exceed the normal force applied so that the COF is always less than one (James 1980). However, these views are incorrect and do not even approximately describe the frictional behaviours of the polymeric and/or elastomeric materials and many rubbers used in the footwear and flooring and floor covering materials (James 1980).

Despite a considerable amount of works have been devoted to studying frictional behaviours, there is no simple model to predict or to calculate friction for a given pair of materials (Dowson 1998). The process of relative sliding between two materials appears to be a simple one, but this development is extremely deceptive. There is no theoretical method of predicting the COF as yet when any two materials are in sliding contact (Suh and Sin 1981; Czichos 1986).

As described in the above, both friction and wear phenomena are the results of extremely complex interactions between the surface and near-surface regions of the two materials. These regions differ from the bulk of the materials in their physical, chemical, and mechanical properties. Furthermore, these properties themselves can change radically as a result of interactions of the surface atoms with their environments and with each other. As a result of this complexity, if it is necessary to know the COF of a particular pair of materials under a specific set of operating conditions, then the safest procedure is to measure it experimentally, under conditions as close to the operating environments as is feasible. In quoting a summarising description by Suh and Sin (1981), the "genesis of friction" may be characterised as

> The coefficient of friction between sliding surfaces is due to the various combined effects of asperity deformation, ploughing by wear particles and hard surface asperities and adhesion between the flat surfaces. The relative contribution of these components depends on the condition of the sliding interface which is affected by the history of sliding, the specific materials used, the surface topography and the environment.

In addition to the above aspects on frictional properties, the following issues also need to be considered:

(1) The COF is not a constant for any particular material that may be similar or dissimilar, sliding against each other under a given set of surface and environmental conditions. In particular, frictional behaviours between the shoe and floor surface are much more complicated than normally expect and combine various sub-mechanisms. As clearly pointed out in the introduction of this book, most research and industry practices on slip and fall safety assessments have simply claimed the results of friction measurements between the shoe and floor surfaces. Most of them were just oversimplified the complex nature and multi-factorial characteristics of frictional and wear behaviours and tribo-physical mechanisms involved. Hence, further studies on the slip resistance measurements between the shoe and floor surface should be based on thorough understanding of the friction and wear activities. The often-posed question such as "What is a COF of this shoe or that floor surface?" actually has no meaning. This means that the concept on the COF itself has not much meaning for evaluating a material performance.

(2) The surface topographies of both coupling surfaces are continuously changing during the process of sliding friction (Kim 1996a, b, c, d, 2004a, b, 2006a, b, 2015a, b, c, d, 2016a, b; Kim and Smith 1998a, b, 1999, 2000, 2001a, b, c, 2003; Kim et al. 2001, 2013; Kim and Nagata 2008a, b). The conditions surrounding an individual asperity and its interaction with an opposing asperity are a crucial point to understand the friction mechanism. Hence, slip resistance results from any friction measurement should clearly demonstrate the details of test settings such as surface features of the shoes and floors, environmental circumstances, product service histories, lubricants used, and floor types such as indoor or outdoor usage. This means that the selection of test conditions for friction measurement requires a great deal of careful considerations.

(3) Most studies and industry practices for the pedestrian slip and fall safety assessments have measured their COFs funder clean and dry surface environments as a first step and then moved into lubricated ones to determine slip

resistance properties of their shoe and/or floor products. This kind of exercise seems to be mainly caused by misleading and/or misunderstanding concepts on the fundamental aspects of contact-sliding friction behaviours between the shoes and floor surfaces and their involved tribo-physical properties (Kim and Smith 2003; Kim 2006a, b). As clearly discussed in the previous chapters, such routine friction measurements significantly involved in changes on the surface features of both shoes and floors. This means that alterations in the topographic conditions of the shoe and floor surfaces are directly related to the frictional behaviours at the sliding interface between them and accordingly slip resistance performance. Specifically, the following issues can be raised:

(a) The surfaces of two pairing bodies can be worn and damaged in a way during repetitive dry sliding friction events so that the rubbed surfaces may create largely different topographic conditions from their initial ones. This result significantly affects the slip resistance functioning and consequently produce misguided information for slip safety judgment between the shoe and floor surface tested.

(b) Without any consideration of the worn surface conditions and their effects on frictional behaviours, further rubbings under the lubricated conditions seem to provide unexpected or distorted outcomes on slip resistance measurements because the altered surface conditions of both bodies may show different responses to the polluted environments. Hence, it can be questioned that

- What effect do the surface changes from the dry friction measurements play in slip resistance properties under the lubricated condition?"
- "How the worn surfaces of the shoe heels affect a short-term and long-term performance of slip resistance?"
- "How the worn surfaces of floor and floor coverings affect a short-term and long-term performance of slip resistance?"

In this context, it can be considered that slip resistance measurements must be counted and assessed between the dry and lubricated conditions and acceptable levels of the COF should be determined, respectively with full details of their surface conditions of both shoes and floors.

(c) Recognising that the lubricated environments show potentially more slip hazardous than the dry ones, but the slip resistance properties between the shoe and floor may have different frictional and wear behaviours. Under the contaminated environments, the friction and wear mechanisms are much more complicated and hard to understand by a simple slip resistance measurement. Hence, the measured slip resistance properties to explain dry sliding friction may provide the basis to explain the mechanisms associated with friction and wear behaviours under the polluted environmental conditions. This approach seems to be a vital step to configure the complex nature of slip resistance properties between the shoe and floor.

5.2.3 The Origins of Friction

5.2.3.1 Introduction

Many early investigators envisaged that major contributions to the frictional force were mainly due to mechanical interactions between the opposing asperities of two rigid sliding surfaces or elastically deforming asperities (Dowson 1998). Each asperity interaction would contribute to the friction force so that the total friction force at any time would be the sum of forces at the individual contacts.

Figure 5.1 illustrates a simple version of this model, often called as the Coulomb model, in which the action of wedge-shaped asperities causes the two surfaces to move apart as they slide from the position A to the position B (Hutchings 1992). It may then be readily shown, by equating the work done by the frictional force to that done against the normal load that θ is equal to tan θ. It is in considering the next phase of the motion, from B to C, however, that a fundamental defect of this model becomes apparent.

The normal load works on the system, and all the potential energy stored in the first phase of the motion (from ***A*** to ***B***) is reversed. No net energy dissipation occurs in the complete cycle, and one must, therefore, conclude that no frictional force should be observable on a microscopic scale if the interaction between real surfaces followed the Coulomb's model exactly.

5.2.3.2 The Amonton's Laws

To mathematically explain the form of the Amonton's laws, the following two assumptions can be made:

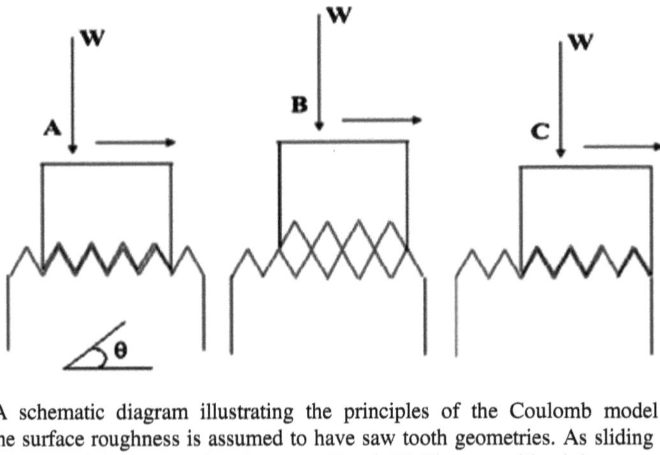

Fig. 5.1 A schematic diagram illustrating the principles of the Coulomb model for sliding friction. The surface roughness is assumed to have saw tooth geometries. As sliding occurs from position ***A*** to ***B***, work is done against the normal load, ***W***. The normal load does an equal amount of work as the surface moves from ***B*** to ***C*** (Hutchings 1992)

(1) During sliding, the resistive force per unit area of contact is constant, so that

$$F = A \times s \quad (5.3)$$

where ***F*** is the friction force, ***A*** is the real area of contact (RCA) and ***s*** is the friction force per unit area (shear strength).

(2) The RCA, ***A***, is proportional to the normal load, ***W***, so that

$$A = q \times W \quad (5.4)$$

where ***q*** is the constant of proportionality.

Eliminating ***A*** from the two equations gives

$$F = q \times s \times W \quad (5.5)$$

If above two assumptions can be justified, then these equations explain the form of Amonton's laws. That is, Eqs. 5.3 and 5.4 indicate that the friction force depends on the RCA and that this is independent of the apparent area, whilst Eq. 5.5 means that friction force is proportional to the normal load. Therefore, it remains to justify the assumptions.

The above assumption (1) can be easily justified. It involves no propositions about the nature of specific friction forces but is simply a statement that any part of the contact areas is statistically representative of the whole.

On the other hand, the above assumption (2) cannot be justified under all circumstances, but can be claimed:

(a) whenever contact is wholly plastic, regardless of the surface topography;
(b) whenever the contacting surfaces have exponential distributions of asperity heights, regardless of the mode of deformation; and
(c) whenever the contacting surfaces have a Gaussian distribution of asperity heights, again regardless of the mode of deformation.

Until the pioneering work of Greenwood and Williamson (1966), which demonstrated the critical importance of surface topography as summarised in the above claims of (b) and (c), Amonton's second law implied that frictional contacts must be plastic. However, almost all rubbing surfaces would simultaneously contain both elastic and plastic contacts, and that it is the generally Gaussian or near-Gaussian surface topographies of most surfaces which explain the wide range of validity of this law.

5.2.3.3 Sliding Friction

Most of the current theories on sliding friction stem from the important work of Bowden and Tabor carried out between the 1930s and 1970s. From the Bowden and Tabor model for sliding friction, in its simplest form, they assumed that the frictional force arose from two main sources: an adhesion force developed in the areas of real contact between two surfaces, and a deformation force needed to plough asperities of the harder surface through the softer one.

From the later developments of the theory, it becomes clear that these two contributions cannot be treated as strictly independent. However, it is convenient and illuminating to consider them separately. The resultant frictional force $\boldsymbol{F}$ is taken to be the sum of two contributing terms, $\boldsymbol{F_{adh}}$ due to adhesion and $\boldsymbol{F_{def}}$ due to deformation. Equation 5.6 shows a relationship amongst the friction, adhesion, and deformation forces:

$$F = F_{adh} + F_{def} \tag{5.6}$$

The following sections further describe the frictional force components of adhesion and deformation with relevant theory models.

5.2.3.4 Adhesion Force

The adhesion component of friction is due to the formation and rupture of interfacial adhesion bonds. When two surfaces are loaded together, they can adhere to some parts of the true contact area to form friction junctions. These junctions must then be broken if relative sliding is to take place. The break would occur at the weakest part of the junction, which may be either at the original interface or in the weaker of the two bodies.

If the break occurs at the original interface, then the interaction has simply subjected the two asperities to a stress cycle, although the accumulation of such cycles may ultimately cause the creation of a wear particle by a fatigue mechanism (Collins and Daniewicz 2006). If the break occurs in the softer one of the two materials, then a fragment of this material will be transferred to the harder counterface (Kim 2015a, b, c, d, 2016a, b). In either case, it is clear that adhesion is one form of surface interactions that would cause frictional resistance.

Simple Adhesion Theory

This theory was the first modern explanation of the existence of friction by Bowden and Tabor (1950). It was developed for ideal elastic-plastic metals. The theory can be summarised as follows:

(1) When surfaces are loaded together, they actually make contact only at the tips of asperities.
(2) Even at low loads, the real contact pressures are so high that the asperity tips of the softer material deform plastically.

(3) This plastic flow causes the total contact area to grow, both by the growth of individual initial contacts and by the initiation of new contacts until the RCA is just sufficient to support the load elastically.

(4) Under these conditions, for an ideal elastic-plastic material

$$W = A \times P_o \tag{5.7}$$

where $\boldsymbol{W}$ is the normal load, $\boldsymbol{A}$ is the real area of contact, and $\boldsymbol{P_o}$ is the yield pressure of the softer one of the two materials. The yield pressure is nearly the same as the hardness, $\boldsymbol{H}$, which can be measured in an indentation test so that this can be rewritten as

$$W = A \times H \tag{5.8}$$

(5) As a result of the severe plastic deformation, the asperity junctions are welded. That is, strong adhesive bonds are formed at the coupling asperities. The specific friction force, $\boldsymbol{s}$, is then simply the force required to cause shear failure of the unit area of asperity junctions, so that

$$F = A \times s \tag{5.3}$$

(6) Neglecting any ploughing contribution, Eqs. 5.8 and 5.3 can be combined to give

$$\frac{F}{W} = \mu = \frac{s}{H} \tag{5.9}$$

It could be seen that the simple adhesion theory provided what was the first theoretical explanation of Amonton's laws mentioned in Sect. 5.2.3.2 that friction was independent of the apparent area of contact.and that friction force was proportional to the normal load.

In the above analysis, however, an ideal elastoplastic material was only considered and the effect of work hardening was ignored. For the case of failure in the softer one of the two materials, it can be taken $\boldsymbol{s}$ equal to $\boldsymbol{k}$, the critical shear stress of this material, so that

$$\mu = \frac{k}{H} \tag{5.10}$$

The ratio $\frac{k}{H}$ is fairly constant for most materials, and Eq. 5.10 indicates why many material pairs have similar COFs despite the fact that the individual values of $\boldsymbol{k}$ and $\boldsymbol{H}$ vary very widely.

Although this simple theory looks attractive, it seems to be inadequate to apply for the shoe-floor sliding friction system from the following two important respects:

(1) The coefficient of friction given by Eq. 5.9 depends only on the mechanical properties of the softer one of the two materials, so it should be expected a particular material to have the same COF against any harder counterface. However, this may not be actually found in the shoe-floor sliding friction mechanism.
(2) Many materials exhibit COFs much higher than those quoted above when their surfaces are free of the normal contaminants and dust (or particles). In the slip resistance measurements between the shoe and floor surface, it is often observed that COFs are much greater than unity on dry and clean surface conditions.

The above main deficiencies lead the Bowden and Tabor model to re-examine some of the assumptions of the simple adhesion theory and to present more realistic descriptions of friction in terms of adhesion force component.

Extension of Simple Adhesion Theory

In the simple adhesion theory, the effects of the normal and tangential loads were considered separately. It was also assumed that the true area of contact was determined by the normal load only (see, Eq. 5.8) and the friction force was taken to be the force required to cause shear over this area (see, Eq. 5.3). However, the very high COFs observed between clean metal surfaces indicate that the true area of contact must be far greater than that predicted by the simple adhesion theory. This was explained when Bowden and Tabor considered the combined effect of the normal and shear stresses (Bowden and Tabor 1950).

5.2.3.5 Deformation Force

Deformation

If no adhesion takes place, then the only alternative interaction which would result in a resistance to motion would be one in which a material was deformed and displaced during relative motions. It is needed to consider two interactions of this type: microscopic and macroscopic interactions (Moore 1975). The microscopic interaction illustrated in Fig. 5.2a requires deformation and displacement of the interlocking surface asperities. And the macroscopic interaction illustrated in Fig. 5.2b means that the asperities of the harder material plough groove in the surface of the softer one.

Other possibilities of the macroscopic type of interaction would involve the ploughing of one or both surfaces of wear particles trapped between them, and macroscopic ploughing of the softer material by the harder one, with the dimensions of the ploughed groove being orders of magnitude greater than those of the asperities on either surface.

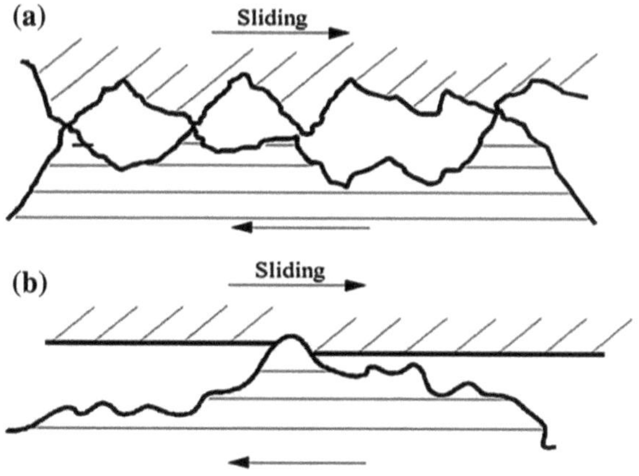

Fig. 5.2 Schematic illustration for sliding interactions between two surfaces: **a** microscopic interaction and **b** macroscopic interaction

Asperity interactions would always be presented and, together with adhesion, would often be the major cause of friction (Roberts 1976). The ploughing contribution may be significant in some cases such as the shoe-floor sliding friction. Its magnitude would depend on the surface roughness and relative harnesses of the two surfaces, size, shape and rigidity of any wear debris and reaction products trapped between them. The following sections further explore how the three processes: adhesion, ploughing and asperity interaction contribute to the friction mechanism with the review of friction theories.

Deformation Theories

In the adhesion theory described above, the normal and yield stresses on a single asperity were assumed to be representative of the stresses on all asperities. In the deformation theories, it is recognised that the normal and shear stresses on the asperities would be changed during the lifetime of a junction.

The physical basis of the analysis is that, in the sliding of macroscopically flat surfaces, the motion is parallel to the interface and the separation of the surfaces remains constant. This must be so to maintain the area of contact at the constant level which would support the constant normal load, and the significant consequence of this is that the contacting asperities must deform to allow movement to continue.

The frictional force (deformation term) due to ploughing of harder asperities through the surface of a softer material may be estimated by considering a simple asperity of idealised shape. If a rigid conical asperity of semi-angle, α, slides over a plane surface as shown in Fig. 5.3, the tangential force needed to displace it would be some flow pressure, which may take as the indentation hardness $\boldsymbol{H}$ of the surface material, multiplied by the cross-sectional area of the groove (Hutchings 1992):

$$F_{def} = Hax = Hx^2 \tan\alpha \qquad (5.11)$$

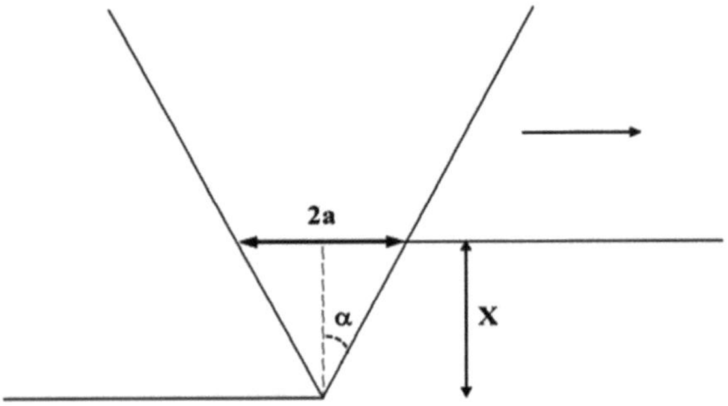

Fig. 5.3 A schematic diagram illustrating the idealised wedge-shaped asperities studied in the plastic interaction theory. The model explains the deformation component of friction, in which a conical asperity of semi-angle **α** indents and slides through the surface of a plastically deforming material (Hutchings 1992)

During sliding events, only the front surface of the asperity is in contact with the softer surface. Hence, the normal load, $\boldsymbol{W_i}$, which is supported by the horizontal projection of the asperity, is given by

$$W = \frac{H\pi a^2}{2} = \frac{1}{2}H\pi x^2 \tan^2 \alpha \tag{5.12}$$

Therefore, the COF due to the ploughing term would be

$$\mu_{def} = \frac{F_{def}}{W} = \left(\frac{2}{\pi}\right) \cot \alpha \tag{5.13}$$

It is a characteristic of all the asperity interaction theories of friction that the energy consumed in plastic deformation increases with the sharpness of asperities. In the case of elastic deformation of asperities, there is no deformation component of friction, but the real contact area of a junction. Hence, the force required shearing it increases with the asperity slope. These two facts are widely accepted that friction between macroscopically smooth surfaces increases with the mean of absolute surface slopes.

The above discussions for the possible forms of frictional behaviour demonstrate why it may never be possible to predict accurately the complex nature of tribo-physical features between the two dissimilar materials, particularly in the case of shoe-floor sliding friction. Recently, research has begun to explain the separation of friction junctions using the ideas of fracture mechanics. This work seems to be very interesting and may well proceed, but it is still in its early stage to applying the concept to the case of shoe-floor sliding friction system.

5.3 Friction Mechanism at the Shoe-Floor Sliding Interface

5.3.1 Introduction

It can be considered that there are main factors contributing to the friction generated between a shoe heel and floor surface in relative motion. Figure 5.4 illustrates principal components of a friction mechanism between a shoe heel and floor surface. It shows a shoe heel sliding over a floor surface at a constant velocity (Fig. 5.4a).

It also demonstrates how the total frictional force developed over a single asperity can be separated into two components: adhesion and a deformation terms, respectively. Figure 5.4b, c presents enlarged views of the circular area in Fig. 5.4a and explains how the two most significant friction components: adhesion and deformation are formed at the sliding interface between the shoe heel and floor surface.

To pull the shoe heel over the floor surface, a frictional force, $\boldsymbol{F}$, is required. This force is essentially the result of adhesion shear, $\boldsymbol{F_a}$, and deformation loses, $\boldsymbol{F_d}$. If it assumes no interaction between the two factors, it may write:

$$F = F_{adhesion} + F_{deformation} \tag{5.14}$$

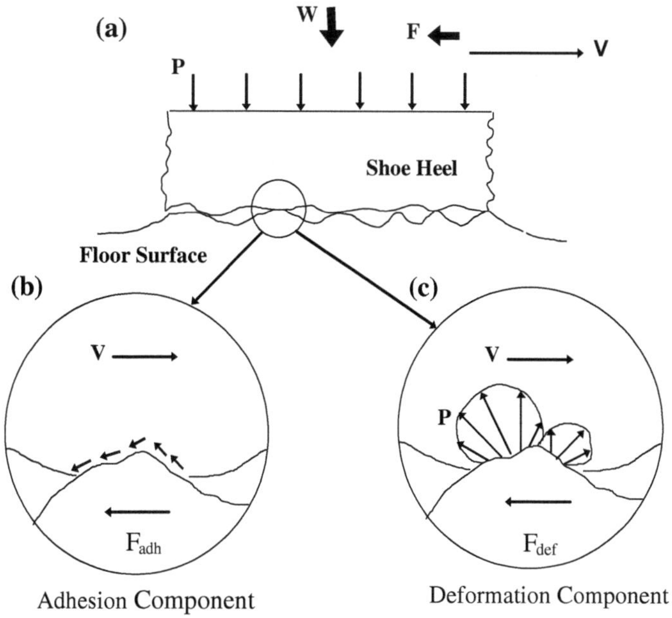

Fig. 5.4 Schematic illustrations of principal components of the friction mechanism between a shoe heel and floor surface. This diagram is based on Moore's model for elastomeric friction (Moore 1972)

Dividing by the load, $\boldsymbol{W}$, the corresponding equation can be written in terms of friction coefficients:

$$F = F_A + F_D \tag{5.15}$$

where the suffixes $\boldsymbol{A}$ and $\boldsymbol{D}$ denote the adhesion and deformation terms, respectively.

Because of the mechanisms responsible for the frictional force components: adhesion and deformation are quite different, their influencing factors such as normal pressure, sliding velocity, surface status, and surface roughness cannot be expected to affect each component in the same manner. Thus, observation of the resultant frictional force, $\boldsymbol{F}$, as a function of any one of these factors does not reveal specific characteristics of the two components, $\boldsymbol{F_A}$ and $\boldsymbol{F_D}$. Following sections discuss each force component in detail.

5.3.2 Adhesion Component

The contact area between a shoe and a floor can be characterised by a draping of the shoe heel's asperities against individual asperities of the floor surface, as shown in Fig. 5.4b.

Due to quasi-elastic behaviours of the heel surface under compression against the floor surface, high normal pressures (approximately 400–1000 psi) are usually developed between them (Kummer and Meyer 1962). These pressures bring the surfaces of the shoe and floor into intimate contact and permit strong molecular forces to be set up between the two bodies. Adhesion is distinctly a surface effect as indicated in Fig. 5.4b, whereas deformation is bulk phenomenon dependent on viscoelastic properties of the shoe heel and sole.

Modern theories of adhesion under a given experimental condition have described adhesion as a thermally activated molecular stick-slip process (Moore 1975). Unlike a hard material, the elastomer structure is composed of flexible chains that are in a constant state of thermal motion. During relative sliding events between an elastomer and a hard material, separate chains in the surface layer attempt to link with molecules in the hard base, thus forming a local junction.

The sliding action causes these bonds to stretch, rupture, and relax before new bonds are made so that effectively the elastomer molecules jump a molecular distance to their new equilibrium position. Thus, a dissipative stick-slip process on a molecular level is fundamentally responsible for adhesion (Schallamach 1953, 1963; Bartenev and El'kin 1965).

As the elastomer moves on, the molecular bond would be sheared again, thereby producing small force components in the local planes of contact. The individual components may be assembled into a resultant force that may be replaced by one component in the plane of movement and another normal to it. Hence, the horizontal component can be described by

$$F_{ai} = s \times A_{ai} \tag{5.16}$$

where A_{ai} is a projection of an enclosed surface that is a real contact area, into the plane of movement and s is shear strength.

The summation of these components over the entire surface (apparent contact area) would give the adhesion component. With the friction continuing, the size of apparent contact area would be enlarged and this would consequently affect the friction result.

$$F_A = s \times \sum_{1}^{n} A_{ai} \tag{5.17}$$

This component also can be increased considerably by pairing an elastomer with a clean smooth surface such as a glass plate for which conditions of s and A_{ai} would be high. According to Kummer and Meyer, COFs can be developed up to 3 and 4 between a rubber block and a clean plate glass at small sliding velocities of 0.1 to 1 in./s. (Kummer and Meyer 1962). On the contrary, such a pairing would be extremely dangerous when lubricant is present.

As mentioned in the above, however, surfaces from the real-world walking environments exhibit both macro- and micro-roughness effects and these in turn directly determine the actual area of contact when a shoe heel is draped over such surfaces under the action of an applied load, W.

The macro-roughness has a mean wavelength, λ, as illustrated by Fig. 5.5. For a given applied load, W, the mean actual pressure, p, on each asperity of the macro-roughness greatly exceeds the apparent or nominal pressure, P_{app}, according to the following relationship:

$$p = \left(\frac{A_{app}}{A_{act}}\right) \times P_{app} \tag{5.18}$$

where the apparent area A_{app} can be visualised as the area of contact for completely smooth surfaces, and A_{act} is the summation of individual areas at the summits of asperities for the actual surface.

Almost all the adhesion theories suggest that F_A is inversely proportional to pressure p so that the existence of a macro-roughness reduces the COF. It is normally desirable that a surface should provide distinct topographic structures to achieve optimum traction properties, and may wonder how this fact can be reconciled with the reduction in adhesion. The solution to this apparent contradiction lies in the distinction between wet and dry surface conditions of sliding friction.

Under the dry surface environment, there is no doubt that the maximum adhesion is attained on smooth surfaces, and since the interfacial area has a maximum value the mechanism of molecular-kinetic bonding is most widespread. In the case

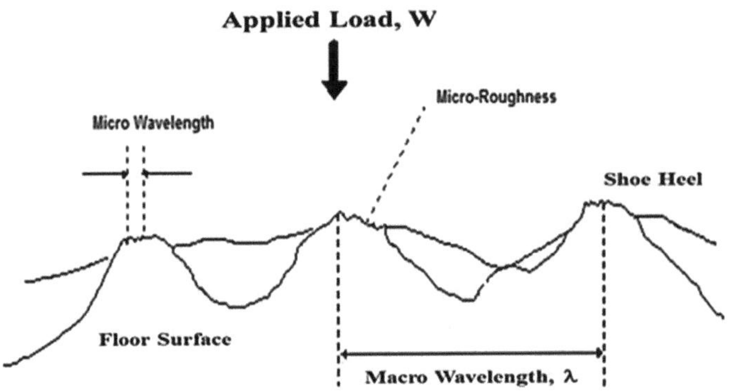

Fig. 5.5 Schematic illustrations for macro- and micro-roughness effects in the contact area

of dry sliding between two rough surfaces, it is recorded that the coefficient of adhesive friction is generally at least twice as large as the deformation contribution (Moore 1972).

Under the wet surface environment, however, an interfacial liquid film is spread uniformly, and it effectively suppresses the roughness effect of surfaces. Thus, the adhesion falls to a very low value. The use of a lubricant between two coupling surfaces on relative motion virtually eliminates the adhesion term and the measured friction force can be attributed solely to the deformation component.

When the rigid surface has a distinct macro-roughness, the voids between asperities act as reservoirs for the liquid under the lubricated surface condition, and the pressure distribution at each asperity summit promotes local drainage effects. As a result, there is a greater probability of suitable conditions existing for adequate adhesion when the surface exhibits a macro-roughness compared with the completely smooth one under the polluted surface condition.

This probability can be greatly enhanced by providing a distinct micro-roughness at asperity peaks. Hence, the combined effects of micro- and macro-roughness under lubricated surface environments are to minimise the decrease in the coefficient of adhesion below the dry value. However, it should be pointed out that many studies in the literature hardly investigated the surface conditions and their effects on slip resistance performance during friction measurements. This means that there is a great deal of uncertainties on the slip resistance measurements such as

(1) how the conditions of initial surface roughness were before any friction test;
(2) what the sizes were;
(3) what types of measurement instrument were used; and
(4) how long test shoe and floor samples were used for friction tests.

The above concerns also should be carefully considered as important factors because physical and mechanical behaviours responsible for the adhesion and deformation forces are widely different to their pairing surface combinations and conditions. Irrespective of whichever theories of adhesion seem most acceptable under a given situation, it becomes clear that the adhesion itself is essentially a contact problem which should be investigated. If the current concept is applied to the shoe-floor sliding friction system, it would be extremely difficult experimentally to determine the exact contact area between the shoe and floor surface. Even most studies in the mechanical engineering fields are greatly limited as the case of single and multiple asperities of idealised shapes such as usually cylinders and spheres.

In practical applications, the effects of randomness of surface texture, physical and mechanical properties, elastohydrodynamic interaction events (for lubricated environmental conditions), and viscoelastic draping behaviours seem to make further complicate the shoe-floor sliding friction system to understand the full extent.

5.3.3 Deformation Component

Figure 5.4c illustrates a mechanism that is responsible for the deformation component, $\boldsymbol{F_d}$. When the shoe heel is moved toward the right, it must climb over the asperity of the floor surface and is subjected to deformation.

The deformation consists of a compression phase on the left side of the floor surface and an expansion phase on the right side. The deformation component of friction can be visualised in detail by showing the pressure distribution about an individual asperity of the surface with two movement modes as shown in Fig. 5.6:

(a) no relative motion at the interface
(b) the presence of relative sliding motion

The following equation appears to be valid for the deformation component of elastomeric friction (Moore 1972):

$$F_D = K \times \left(\frac{p}{E}\right)^n \times \tan\alpha \tag{5.19}$$

where $\boldsymbol{F_D}$ is the deformation force, $\boldsymbol{p}$ is the nominal pressure and $\boldsymbol{K}$ is a constant. The **tan** $\boldsymbol{\alpha}$ is the ratio of energy dissipated (whether in the stretch and rupture cycle of individual bonds in the molecular adhesion model, or in the deformation and recovery cycle associated with deformation) to energy stored per cycle.

From Eq. 5.19, it can be observed that adjustment of the nominal pressure, $\boldsymbol{p}$, in a practical design problem necessitates a compromise between the magnitudes of adhesion and deformation force components, and determination of the optimum pressure depends particularly on the surface roughness of the base material and on the presence or absence of an effective interfacial lubricant.

The deformation component and its magnitude have evoked considerable interest and controversy amongst researchers. From a fall safety point of view, however, most of the walkway and floor surfaces for the pedestrian are rough enough so that the deformation force seems to dominate the frictional force.

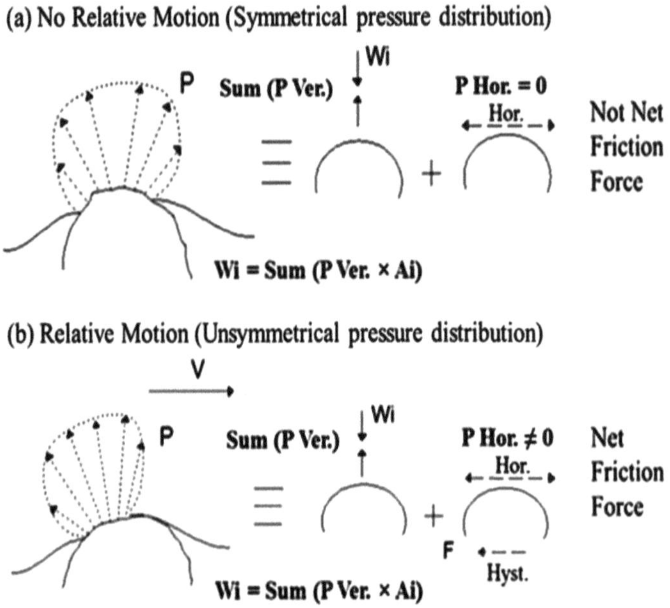

Fig. 5.6 Schematic illustrations for the physical interpretation of the deformation component of friction (Moore 1972)

The result of high frictional force is, of course, wear or abrasion development. In the case of polymer or elastomer material, three distinct mechanisms of wear formation can be identified, depending on the nature of surface texture (Moore 1972).

(a) For very sharp surfaces, abrasive wear produces severe degradation of the elastomer surface, corresponding to a high coefficient of friction.
(b) When the surface asperities are rounded rather than sharp, fatigue wear predominate, and it constitutes a relatively mild form of surface deterioration.
(c) On smooth surfaces, there is evidence that wear development by roll formation occurs on the surface of the elastomer, and it is characteristically accompanied by a high coefficient of sliding friction.

In each application, all the above three forms of wear progress may coexist simultaneously. Hence, it becomes evident that wear behaviours between the shoe heel and floor surface during repetitive ambulation should be investigated with friction measurements. However, these important aspects of wear behaviours and wear effects on slip resistance performance have not been explored in the fall safety research area.

Following sections briefly deal with basic information on wear problems of the shoe and floor surface during sliding friction measurements and suggest model developments for the slip resistance analysis between the shoe heels and floor surfaces.

5.4 Wear Mechanism

5.4.1 Introduction

Similar to the frictional mechanism, wear behaviours of the materials also show a very complicated phenomenon in which various mechanisms and influencing multi-factors are involved.

A great step forward in the understanding of wear phenomena was the classification of wear mechanisms given by Burwell in the 1950s (Burwell 1957), according to which wear mechanisms might be divided into three broad general classes under the headings of abrasion, adhesion, and fatigue.

An increasing number of studies have devoted to understanding wear behaviours for the industrial and manufacturing products and reported that wear event is the removal of materials from interacting surfaces in relative motion, and the result from various interactive processes (Czichos 1982; Briscoe 1981; Rigney 1980; Habig 1980).

In quoting a summarising description on the wear processes according to Suh (1982),

> Wear of materials occurs by many different mechanisms depending on the materials, environmental and operating conditions, and geometry of wearing bodies. These wear mechanisms may be classified into two groups: those primarily dominated by mechanical behaviours of solids and those primarily dominated by chemical behaviours of materials. What determine the dominant wear behaviours are mechanical properties, chemical stability of materials, temperature, and operating conditions.

In Tabor's recent critical synoptic review of wear behaviours (Tabor 1979), he simply divided the wear processes into three groups:

(1) Wear arises primarily from adhesion between the sliding surfaces,
(2) Deriving primarily from non-adhesive processes, and
(3) The interaction between adhesive and non-adhesive processes to produce a type of wear that seems to have characteristics of its own.

The way in which these mechanisms interact with one another depends sensitively on the specific operating conditions. In addition, the frictional process itself can produce profound structural changes and modifications on the physical and chemical properties of the sliding surfaces. Consequently, unless a single wear process dominates, these surface changes and complex interactions must necessarily make wear predictions extremely difficult and elusive (Tabor 1979).

5.4.2 Main Considerations

Elastomers and polymers are much more compliant than metals or ceramics, with values of elastic modulus typically one tenth or even less. Because of their strengths are also much lower than the metallic or ceramic counterface, nearly all the deformation due to contact or sliding takes place within the elastomers when they slide against to the rigid bodies.

In this situation, surface finishes of the hard counterface have strong influences on the resulting wear mechanism. For example,

(1) If the counterface is smooth, then wear may result from adhesion between the surfaces, and involve deformation only in the surface layers of the elastomer.
(2) On the other hand, if the counterface is rough then its asperities may cause deformation in the elastomer to a significant depth. Wear then results either from abrasion associated with plastic deformation of the elastomer or from fatigue crack growth in the deformed region (Tanaka 1995).

The above two classes of wear mechanisms, involving surface and subsurface deformation, are termed as interfacial and cohesive wear processes, respectively. The level of counterface roughness at which the transition from interfacial to cohesive wear mechanisms occurs depends on the nature of elastomers but corresponds typically to the surface roughness. Hence, the importance of surface roughness of both sliding bodies becomes evident.

In order to investigate the process of wear development, the whole complex of the generation of loose wear products must be considered. The chain of events that lead to the creation of wear particles and material removals from a given tribo-physical system can be initiated by two broad classes of tribological processes as summarised in Fig. 5.7.

(1) Stress interactions: These are due to the combined action of load and frictional forces and lead to wear processes, described broadly as surface fatigue and abrasion.
(2) Material interactions: These are due to intermolecular forces either between the interacting solid bodies or between the interacting solid bodies and the environmental atmosphere (and/or the interfacial medium) and lead to wear processes, described broadly as adhesion.

Figure 5.8 suggests a tribo-physical system between a shoe heel and floor surface during sliding friction events. As shown in Fig. 5.8, there are many factors involved in the sliding interfaces between the shoe and floor during the sliding friction measures. Hence, any slip safety assessments should be based on fundamental understanding and thorough investigation of friction and wear phenomena of the shoes and floors and their interactive behaviours on slip resistance performance.

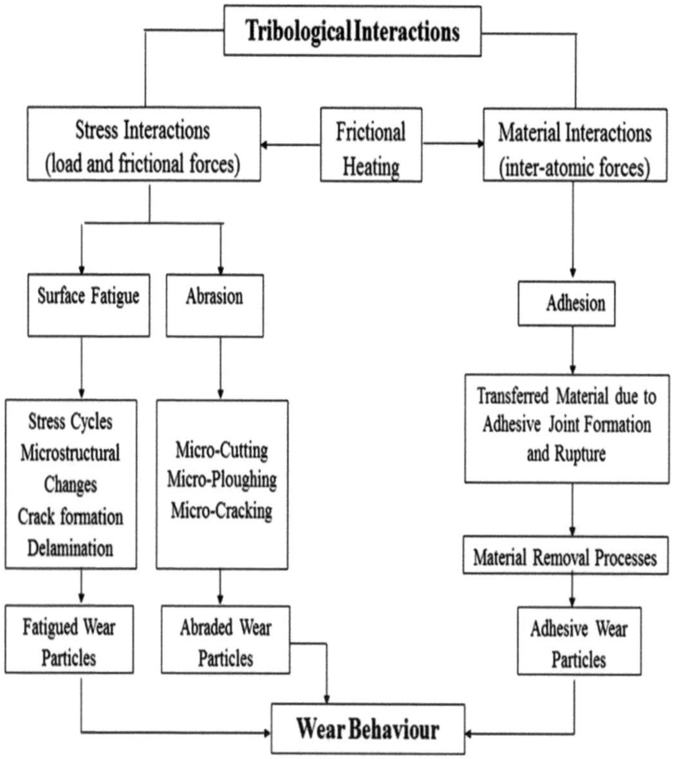

Fig. 5.7 A schematic diagram for tribological interactions and wear mechanisms

5.5 Wear Model for the Shoe-Floor Sliding Friction System

As discussed in the above sections, friction measurements to assessing slip resistance functioning between the shoe heel and floor surface seem not provide intrinsic properties but depend on so many influencing factors.

In any given walking environment, all the possible factors from the proposed 'tribo-physical system' should be considered. Thus, measured quantities of the friction and wear, e.g. COFs and wear rates (mild or severe) are significantly depended upon the following basic groups of parameters.

(a) Structures of the shoe-floor tribosystem, i.e. material properties of the shoe and floor and relevant properties of the system's components.
(b) Operating variables, including normal load, frictional load, kinematics, velocity, operating duration, and contact strike angle.
(c) Interactions between the shoe-floor trib-physical system components.

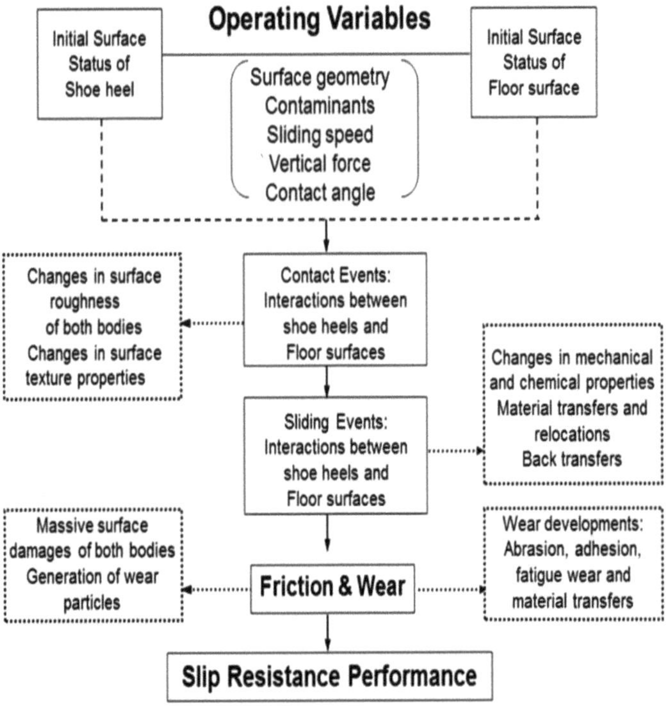

Fig. 5.8 A schematic diagram for a presumed tribo-physical system between a shoe heel and floor surface during sliding friction events

For the close observation, the wear processes need to be broken down into several sub-elements. Figure 5.9 shows an example for a whole wear cycle of a shoe surface from an initial heel contact to wear formation on the heel surface during repetitive sliding friction against a floor surface.

As shown in Fig. 5.9, since all the steps leading to wear development, are mainly caused by the forces acting on real areas of contact (RCAs), macroscopic operating conditions such as shoe tread patterns, shoe material types, floor surface finishes, floor material types, and environments cannot define the wear process by themselves.

For example, particular micro- and macro-tread patterns of the shoe sole and/or heel surface may help to determine the RCAs between the shoe and floor surfaces under a range of walking conditions. The actual forces working on the RCAs then control the subsurface stress fields and strain distributions, which result in the accumulation of damages leading eventually to the removal of wear products.

An important point in this process is that the formation of wear products is accompanied by gradual and/or sometimes abrupt changes of geometry in the heel surface. Thus, a feedback loop as shown in Fig. 5.9 is considered during the repetitive sliding events. As a result, worn surfaces of the shoe heel may affect the contact sliding conditions and eventually slip resistance results.

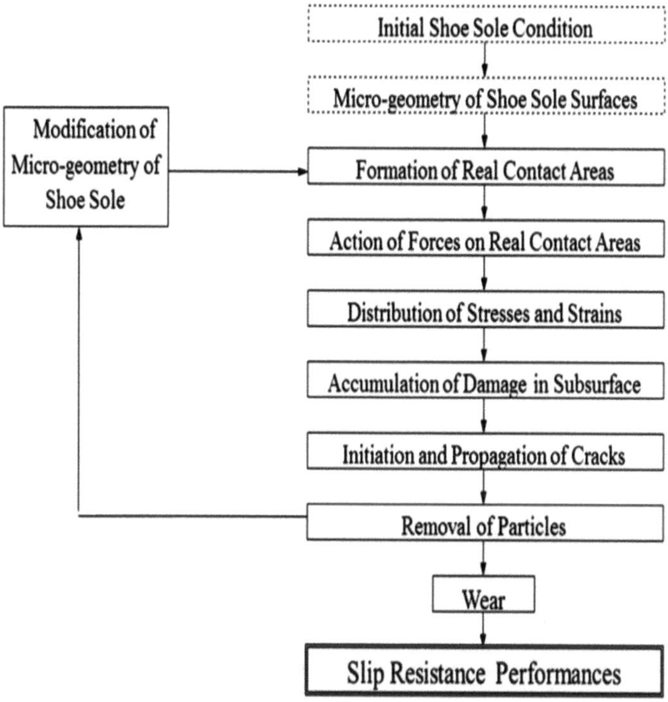

Fig. 5.9 A schematic diagram for the breakdown of a whole wear cycle into elemental processes for a shoe surface during repetitive sliding events against a floor surface

As described in the above on wear processes, micro-topographies of the sliding interface between the shoe heel and floor surfaces are not a given property but continuously change during the repeated friction process. This causes wide differences in the slip resistance results and wear rates often experienced between macroscopically similar types of the shoes and floors during their slip resistance measurements.

The overall procedure for an organised analysis on the shoe-floor tribo-physical system is proposed in Fig. 5.10 as a simplified diagram format. In considering technical entities at which friction and wear processes occur, the material components and substances of both shoe and floor surfaces directly participating in the friction and wear processes have firstly to be identified and conceptually separated from the other parts of the technical units considered. These components are a part of the entire structure for the tribo-physical system of the shoe-floor sliding interface, designated as (1)–(4) in Fig. 5.10.

In addition to the structural elements, the operating variables also need to be identified. Through the action of operating variables on the structural elements of the shoe-floor sliding friction system, tribo-physical interactions occur which directly lead to changes of the topographic characteristics of both bodies from the shoe-floor system components as well as to friction-induced energy losses, and wear-induced material damages.

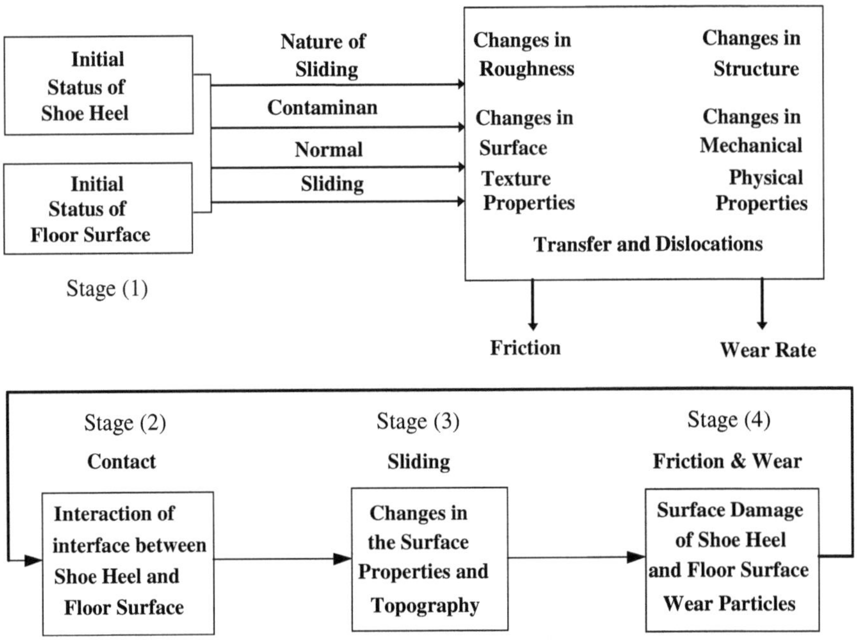

Fig. 5.10 A schematic diagram for basic parameter groups of a tribo-physical system at the shoe-floor sliding interface. Each diagram demonstrates a whole course and factors which affect the friction and wear mechanisms of the sliding interface between the shoe and floor surface (Kim 2006a, b)

5.6 Chapter Summary

Factors that are influencing slip resistance properties between the shoe heel and the floor surface are mainly classified into two components in terms of frictional force. The contribution of the resultant frictional force, $\boldsymbol{F}$, as a function of both force components totally depends on the various factors such as normal pressure, sliding force, sliding speed, and surface texture.

Amongst all these factors mentioned, however, the surface characteristics and their interactions are eventually one of the major reasons to affect frictional and wear behaviours of both shoe and floor and their sliding interface between them. Because the surfaces of almost all solid materials are far from smooth and even the shoes and pedestrian floorings seem to be coarser and rougher than the shoe and floor specimens tested in the laboratory conditions. It is, therefore, necessary to investigate the variations of surface features with their friction and wear behaviours during slip resistance measurements.

The above descriptions of the possible forms of frictional and wear behaviours clearly demonstrated why it might never be possible to predict accurately their tribo-physical characteristics between two solid materials, particularly in the case of sliding friction between the shoe heel and floor surface. A systematic approach was discussed in which friction and wear processes between the shoe heel and floor surface during the prolonged sliding events were broken down into several elemental courses forming a feedback loop and a newly developed model for the wear mechanisms between the shoe and floor surface was suggested.

It is considered that theoretical developments for the shoe-floor sliding surfaces and their tribo-physical characteristics have not yet reached a mature stage where it may be possible to quantitatively predict friction and wear behaviours from known surface characteristics, but this seems to be a useful diagnostic tool and a step forward to identifying the complex issues.

In this chapter, therefore, the origin of frictional forces was examined and related tribo-physical features such as friction induced wear evolutions at the sliding interface between the shoe and floor surface were extensively explored. However, the discussion was mainly focused on understanding for the friction and wear behaviours of the shoe and floor surface under unlubricated environments in sliding motions as a first instance.

A wide variety of conditions causes wear activities with many sub-mechanisms such as adhesion, abrasion, corrosion, and fatigue contributing to the damage caused. There are situations where one type changes to another, or where two or more mechanisms operate together. In any particular instance of wear development, one may have any of these mechanisms operating either individually or in combination.

In the latter case, the situation is even more complicated because of interactions between the multi- faceted mechanisms involved. Therefore, the wear behaviours of both shoes and floors and their interactive effects should be investigated for the analysis of slip resistance properties with friction measurements. However, wear phenomena between the shoe heel and floor surface are rarely considered in the pedestrian fall safety study until now.

References

Bartenev, G. M., & El'kin, A. I. (1965). Friction properties of high elastic materials. *Wear, 8*(1), 8–21.

Bhushana, B. (2003). Adhesion and stiction: Mechanisms, measurement techniques, and methods for reduction. *Journal of Vacuum Science & Technology, 21*(6), 2262–2296.

Bowden, F. P., & Tabor, D. (1950). *The friction and lubrication of solids*. Oxford, UK: Clarendon Press.

Briscoe, B. J. (1981). Wear of polymers: An essay on fundamental aspects. *Tribology International, 14*(4), 231–243.

Burwell, J. T. (1957). Survey of possible wear mechanisms. *Wear, 1*(2), 119–141.

Collins, J. A., & Daniewicz, S. R. (2006). Failure modes: Performance and service requirements for metals. In M. Kutz (Ed.), Mechanical engineers' handbook: Materials and mechanical design (Vol. 1, 3rd Ed., Chapter 26, pp. 860–923). New York: Wiley.

Czichos, H. (1974). Failure criteria in thin film lubrication: The concept of a failure surface. *Tribology, 7*(1), 14–20.

Czichos, H. (1982). *Friction and wear of materials, components and constructions*. Grafenau: Expert-Verlag (in German).

Czichos, H. (1986). Introduction to friction and wear. In K. Friedrich (Ed.), *Friction and wear of polymer composites*. Amsterdam, The Netherlands: Elsevier.
Dowson, D. (1998). *History of tribology* (2nd Ed.). Professional Engineering Publishing (Distributed by ASTM).
Greenwood, J. A., & Williamson, J. B. P. (1966). Contact of nominally flat rough surfaces. *Proceedings of Royal Society, London, A295*(1442), 300–319.
Habig, K. H. (1980). *Wear and hardness of materials*. Munich: Hanser Verlag (in German).
Hutchings, I. M. (1992). *Tribology—Friction and wear of engineering materials*. Metallurgy and Materials Science Series. London: Edward Arnold.
James, D. I. (1980). A broader look at pedestrian friction. *Rubber Chemistry and Technology, 53* (3), 512–541.
Kim, I. J. (1996a). Tribological concepts for the investigation of the pedestrian slipping and falling accidents—Part I. International Occupational Injury Symposium (p. 101), February, Sydney, Australia.
Kim, I. J. (1996b). Tribological approach for the analysis of pedestrian slip hazard—II. Proceedings of the '96 Spring Conference of Korean Institute of Industrial Engineers (pp. 279–285), April, Seoul, Korea.
Kim, I. J. (1996c). Microscopic investigation to analyze the slip resistance of shoes. Proceedings of the 4th Pan Pacific Conference on Occupational Ergonomics (pp. 68–73), November, Taiwan, ROC.
Kim, I. J. (1996d). Microscopic observation of shoe heels for pedestrian slip hazard investigation. Proceedings of the 1st Annual International Conference on Industrial Application and Practice (pp. 243–250), December, Texas, U.S.A.
Kim, I. J. (2004a). Development of a new analyzing model for quantifying pedestrian slip resistance characteristics: Part I—Basic concepts and theories. *Industrial Journal of Industrial Ergonomics, 33*(5), 395–401.
Kim, I. J. (2004b). Development of a new analyzing model for quantifying pedestrian slip resistance characteristics: Part II—Experiments and validations. *Industrial Journal of Industrial Ergonomics, 33*(5), 403–414.
Kim, I. J. (2006a). The current hiatus in fall safety measures. In W. Karwowski (Ed.), *International encyclopedia of ergonomics and human factors-2005* (pp. 2572–2576). New York, USA: Taylor & Francis Group, LLC.
Kim, I. J. (2006b). A new paradigm for characterizing slip resistance properties. In W. Karwowski (Ed.), *International encyclopedia of ergonomics and human factors-2005* (pp. 2735–2740). New York, USA: Taylor & Francis Group, LLC.
Kim, I. J. (2015a). Practical design search for optimal floor surface finishes to prevent fall incidents. In B. Evans (Ed.), *Accidental Falls: Risk Factors, Prevention Strategies and Long-Term Outcomes* (Chapter 5, pp. 80–103). Hauppauge, NY, USA: Nova Science Publishers, Inc.
Kim, I. J. (2015b). Slip-resistance measurements for assessing pedestrian falls: Facts and fallacies. In B. Evans (Ed.), *Accidental Falls: Risk Factors, Prevention Strategies and Long-Term Outcomes* (Chapter 6, pp. 105–125). Hauppauge, NY, USA: Nova Science Publishers, Inc.
Kim, I. J. (2015c). Wear observation of shoe surfaces: Application for slip and fall safety assessments. *Tribology Transactions, 58*(3), 407–417.
Kim, I. J. (2015d). Research challenges on slip-resistance measurements for assessing pedestrian fall incidents. *Journal of Ergonomics, 5*(3). doi:10.4172/2165-7556.1000e142
Kim, I. J. (2016a). A study on wear development of floor surfaces: Impact on pedestrian walkway slip-resistance performance. *Tribology International, 95,* 316–323.
Kim, I. J. (2016b). Identifying shoe wear mechanisms and associated tribological characteristics: The importance for slip resistance evaluation. *Wear, 360–361,* 77–86.
Kim, I. J., Hsiao, H., & Simeonov, P. (2013). Functional levels of floor surface roughness for the prevention of slips and falls: Clean-and-dry and soapsuds-covered wet surfaces. *Applied Ergonomics, 44*(1), 58–64.

Kim, I. J., & Nagata, H. (2008a). Nature of the shoe wear: Its uniqueness, complexity and effects on slip resistance properties. In *Contemporary ergonomics 2008* (Vol. 15, pp. 728–734). London: Taylor & Francis.
Kim, I. J., & Nagata, H. (2008b). Research on slip resistance measurements—A new challenge. *Industry Health, 46*(1), 66–76.
Kim, I. J., & Smith, R. (1998a). A study of the comparative geometry mating between the surfaces of the shoe and floor in pedestrian slip resistance measurements. The 5th Pan-Pacific Conference on Occupational Ergonomics (pp. 34–37), July, Kitakyushu, Japan.
Kim, I. J., & Smith, R. (1998b). Tribological characterization of the frictional force component in pedestrian slip resistance measurements. Third World Congress of Biomechanics (WCB '98), August, Hokkaido, Japan.
Kim, I. J., & Smith, R. (1999). The relationship between wear, surface topography characteristics and coefficient of friction as a means of assessing the slip hazards. 2nd Asia-Pacific Conference on Industrial Engineering and Management Systems (APIEMS'99) (pp. 155–161), October, Ashikaga, Japan.
Kim, I. J., & Smith, R. (2000). Observation of the floor surface topography changes in pedestrian slip resistance measurements. *Industrial Journal of Industrial Ergonomics, 26*(6), 581–601.
Kim, I. J., & Smith, R. (2001a). A critical analysis on the friction measuring concept for slip resistance evaluation. ASTM Symposium on the Metrology of Pedestrian Locomotion and Slip Resistance (pp. 1–14), June, ASTM Headquarters, West Conshodocken, Pennsylvania, USA.
Kim, I. J., & Smith, R. (2001b). A study for characterising topography changes of shoe surfaces in the early stage of slip resistance measurements—Bearing Area Curve. 6th Pan-Pacific Conference on Occupational Ergonomics (pp. 299–303), August, Beijing, P. R. China.
Kim, I. J., & Smith, R. (2001c). Three-dimensional analysis of floor surface wear during slip resistance measurements. 6th Pan-Pacific Conference on Occupational Ergonomics (pp. 304–308), August, Beijing, P. R. China.
Kim, I. J., & Smith, R. (2003). A critical analysis of the relationship between shoe-floor wear and pedestrian/walkway slip resistance. In M. I. Marpet & M. A. Sapienza (Eds.), *Metrology of pedestrian locomotion and slip resistance*. Philadelphia, USA: American Society of Testing and Materials, Special Technical Publication 1424, ASTM International.
Kim, I. J., Smith, R., & Nagata, H. (2001). Microscopic observations of the progressive wear on the shoe surfaces which affect the slip resistance characteristics. *Industrial Journal of Industrial Ergonomics, 28*(1), 17–29.
Kummer H. W., & Meyer, W. E. (1962). Measurement of skid resistance. Symposium on Skid Resistance (pp. 3–28). ASTM Special Technical Publication, No. 326.
Moore, D. F. (1972). The friction and lubrication of elastomers. International Series of Monographs on Materials Science and Technology (Vol. 9). Headington Hill Hall, Oxford: Pergamon Press Ltd.
Moore, D. F. (1975). *Principles and applications of tribology*. Pergamon International Library of Science, Technology, Engineering and Social Studies, Exeter, UK.
OECD. (1968). *Friction, lubrication and wear—Terms and definitions*. Paris: Research Group on the Wear of Engineering Materials.
Rigney, D. A. (1980). *Fundamentals of friction and wear of materials*. ASM Materials Science Seminar, 4–5 October 1980, Pittsburgh, Pennsylvania, Seminar Committee of the Materials Science Division of the American Society for Metals, in cooperation with the Metallurgical Society of AIME, USA.
Roberts, A. D. (1976). Theories of dry rubber friction. *Tribology International, 9*(2), 75–81.
Schallamach, A. (1953). The velocity and temperature dependence of rubber friction. *Proceedings of the Physical Society, Section B, 66*(5), 386–392.
Schallamach, A. (1963). A theory of dynamic rubber friction. *Wear, 6*(5), 375–382.
Suh, N. P. (1982). Surface interactions. In P. B. Senholzi (Ed.), *Tribological technology* (Vol. 1, pp. 37–208). Den Haag-Boston-London: Martinus Nijhoff.

Suh, N. P., & Sin, H. C. (1981). The genesis of friction. *Wear, 69*(1), 91–114.
Tabor, D. (1979). *Proceedings of international conference on wear materials* (p. 1). New York: American Society of Mechanical Engineers.
Tanaka, K. (1995). Some interesting problems that remain unsolved in my work on polymer tribology. *Tribology International, 28*(1), 19–22.

Chapter 6
Surface Measurement and Analysis

6.1 Introduction

Understanding the movement of one solid surface over another seems to be fundamentally important for assessing the shoe-floor sliding friction behaviour as an essential prerequisite. To observe tribo-physical characteristics of the shoe and floor and its interactions at the sliding interface, it would be necessary to identify how two surfaces interrelate when they are loaded together. This is one of the primary purposes of this chapter to develop such understanding. The subject of this matter would lead to developing clear ideas of what surface means in the context of comprehending the shoe-floor friction mechanism and its related tribo-physical characteristics.

Traditionally, we may think of the surface of a solid as a geometrical boundary between the solid and its environment. For one of the main purposes of this book, however, such a definition is too limited so that it must prove a good deal into the body of a solid in order to understand how its surface behaves.

Tribological behaviours are influenced by not only physical, chemical, and mechanical properties of interacting materials but also the near-surface ones (Holmberg and Matthews 2000). In order to study the pedestrian slip resistance properties, therefore, it shall implicitly consider both surfaces of shoes and the floors simultaneously including the nature of surface layers and sub-surface behaviours of the two interacting bodies. It shall be included the physical properties of surface layers as well as geometric characteristics and interactions of actual surface profiles of both bodies. These characteristics largely depend on the bulk properties of solid surfaces, the methods by which the surfaces are produced, and the nature of environments around the surfaces. Thus, optical behaviours of both bodies seem to be governed by the smoothness of solid surfaces, the character of surface layers and the extent to which one surface slides against the other one to begin a relative movement.

I.-J. Kim, *Pedestrian Fall Safety Assessments*,
DOI 10.1007/978-3-319-56242-1_6

Primarily, because of their high strength and toughness, metals have been the most common tribological materials for many years (Rabinowicz 1965; Arnell et al. 1991; Dowson 1998). For this reason, their tribo-physical behaviours have been studied in greater detail and are consequently more fully understood than that of other materials. Because of huge ranges of shoe and floor materials available from the market and diverse characteristics of tribo-physical properties, it is very difficult to examine their friction behaviours directly. Therefore, studying on the nature of metal surface as a starting point would be beneficial to approach the complex shoe-floor friction problem.

6.2 Nature of Surfaces and Their Contact Mechanism

6.2.1 *Fundamental Concepts*

Almost all surfaces are rough on a microscopic scale and the variation in surface profiles can be represented by a random arrangement of peaks and valleys, as shown in Fig. 6.1a. When two surfaces are in contact, it can be assumed that they touch only at tiny discrete areas where their highest asperities are in contact, as shown in Fig. 6.1b. Thus, in general, the real area of contact (RCA) is only a small percentage of the nominal contact area (NCA). The local pressure at the contact regions is then high enough to cause plastic deformation of the asperities even at the lightest load (Stachowiak and Batchelor 2005; Kim 2006a, 2015a, c, 2016a, b; Kim and Nagata 2008a, b; Kim et al. 2013).

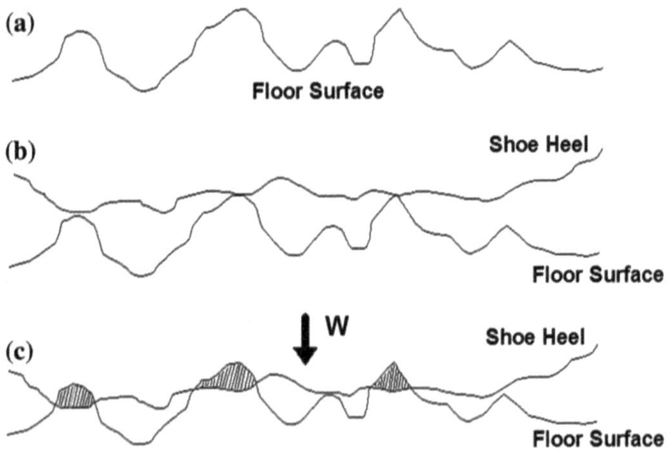

Fig. 6.1 Brief schematic description of the surface nature between a shoe heel and floor surface and their interaction during static contact

In a case of the shoe-floor contact sliding system, the general state of contact spots could be described as Fig. 6.1c. Its contact mechanism seems to be elasto-plastic state and have an interlocking status. The interlocking mechanism seems to be very complex and be governed by a number of factors such as the shapes, sizes and distributions of asperities of both coupling surfaces of the shoe heel and floor, material properties, normal loads, environmental conditions (dry and/or lubricated), and sliding speeds under which the contact occur. In addition, when sliding occurs, the problem would be much harder to approach than the static contact situation because the sliding would accompany wear development with friction processes.

When two such different surfaces move over each other, wear development seems to occur and damage to one or both surfaces, generally involving progressive loss of materials. In most cases, wear is detrimental, leading to increased clearances between the moving components, unwanted freedom of movement and loss of precision, often vibration, increased mechanical loading and more rapid wear, and sometimes fatigue failure (Johnson 1985). The wear loss by relatively small amounts of material can be enough to cause complete failure of large and complex machines (Ahmed and Hadfield 1999; Lim et al. 1999; Nakajima et al. 2000).

The interaction of two solid bodies occurring in the frictional area can only be understood properly if the surface geometry of pairing parts is taken into account. Hence, development of a contact theory model between the shoe and floor surfaces should be based on the classical solutions of problems for elastic and plastic types of contact and on the results of research into the surface quality and analysis. This needs to involve principal understanding on the elastic and plastic flows of contacting surfaces. Both friction and wear events are due to the forces which arise from the contact of solid bodies in relative motion. Therefore, thorough comprehension on the contact mechanism should be progressed for the better approaches to friction and wear problems between the shoe and floor surfaces.

6.2.2 Contact Mechanism Between Two Surfaces

When two nominally-plane and parallel surfaces are brought together, contact seems to initially occur at only a few highest points. As the normal load is increased, the surfaces move closer together and a larger number of the higher areas or asperities on the two surfaces come into contact (Johnson 1985). Since these asperities provide the only points at which the surfaces touch, they are responsible for supporting the normal load on the surface and for generating any frictional forces, which act between them.

Solid bodies subjected to an increasing load deform elastically until the stress reaches a limiting value, known as the yield stress, and at higher stress, they then deform plastically in which the material is permanently deformed (Flinn and Trojan 1975; Kumagai et al. 1978; Dieter 1986). In most contact situations, some asperities are deformed elastically, whilst others are deformed plastically. The loads would induce a generally elastic deformation of the solid bodies but local plastic deformation may take place at the tips of the asperities where the actual contact occurs (Arnell et al. 1991).

Based on the above concept, it can be assumed that contact situations at the sliding interface between the shoe heel and floor surface are a combination of elastic and plastic deformations because of their unique pairing material properties. This chapter is mainly focused on two solid bodies, which are in contact sliding movements because the applied loading and the nature of transitions from elastic to plastic behaviours are of great importance to understand the shoe-floor sliding friction mechanism.

It becomes evident that knowledge of the way in which the asperities of two surfaces interact under varying loads is, therefore, essential information to understand frictional behaviours and their related tribo-physical characteristics such as surface features and their alterations, wear developments of the shoe and floor surfaces, and their interactive effects on slip resistance performance between the surfaces of both bodies. Following sections further describe these issues in detail and formulate their relationships.

6.2.3 Simple Theory of Rough Surface Contact

A surface seems to be considered as an array of spherical asperities (peaks) whose heights follow some particular distribution laws. Consider that the rough surface consists of such an array of spherical asperities, which have the same height and radius and each asperity deform independently of all the others, in contact with a smooth rigid plane (see, Fig. 6.2).

Any load, $\boldsymbol{W}$, applied to such a system would be equally divided between the asperities each of which has a load $\boldsymbol{W_1}$ such that (Arnell et al. 1991)

$$\boldsymbol{W} = \sum \boldsymbol{W_1} \tag{6.1}$$

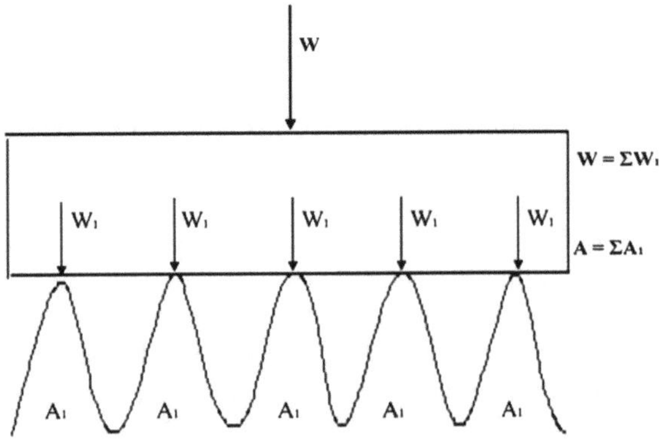

Fig. 6.2 A schematic diagram model of the contact between a smooth surface and a rough surface having asperities of the same height

Each asperity would bear a same fraction of the total normal load, and each would contribute the same area to the total area of contact (TCA). By summing the contributions from all the asperities over the whole area of contact, the total real area of contact (TRCA), $\boldsymbol{A}$, is the sum of each of the discrete areas $\boldsymbol{A}_i$ and $\boldsymbol{A}_i$ would be given by πa_i^2 where a_i is the radius of each circular contact spot.

For the case of purely elastic contact,

$$A \propto W^{\frac{2}{3}} \tag{6.2}$$

$$A_i = \pi a_i^2 W_i^{\frac{2}{3}} \tag{6.3}$$

And, for perfectly plastic behaviour of the asperities,

$$A \propto W \tag{6.4}$$

It has been known that the mean contact pressure of each contact for complete plastic behaviour is the material hardness (Johnson 1985). Thus,

$$\frac{W_1}{\pi a_1^2} = H \tag{6.5}$$

$$\pi a_1^2 = A_1 = \frac{W_1}{H} \tag{6.6}$$

The above two results note that for elastic deformation the TRCA of contact is proportional to $W^{2/3}$, whereas for plastic deformation it is proportional to $\boldsymbol{W}$. For most real surfaces, however, the asperities would have different heights and radius. As a result, both heights and radii of the surface irregularities would be statistically distributed. Hence, at the lightest loads, contact would only occur at the very highest ones (see, Fig. 6.3).

As the load on the surface is increased, not only would the discrete areas of these initial contact increase, but more new contacts would be created at the lower asperities and started to carry some loads. From the information on the asperity height distribution, the actual number of asperities in contact can be defined at any degree of compression of the surface texture, which has occurred. Although the method of treating this contact problem is beyond the scope of this book, the applications of such a theory seem to be a worthwhile concept to develop physical understandings on the shoe-floor contact mechanism.

Following is a summary of the most important observations on the contact behaviours between two rough surfaces. Since many surfaces have nearly Gaussian asperity-height distributions, some conclusions for the contact mechanisms between such interacting surfaces are stated in the following (Arnell et al. 1991):

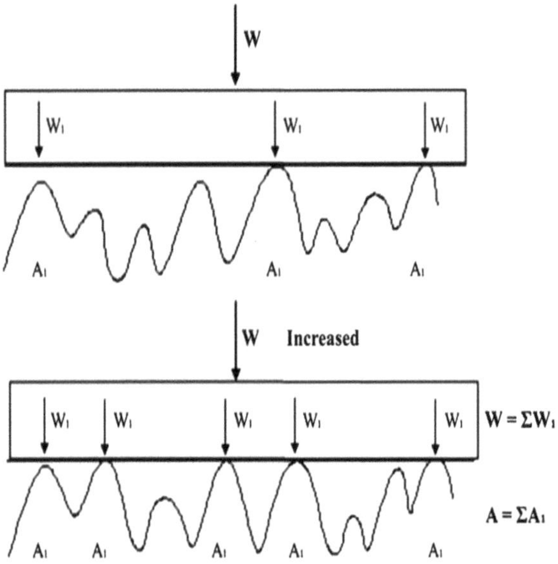

Fig. 6.3 A schematic diagram model of the contact between a smooth surface and a rough surface having asperities of varying heights

(1) As the load increases and the smooth surface compresses the rough one, the mean area of a contact point remains constant. That is, the TCA, $\boldsymbol{A}$, divided by the number of contacts, $\boldsymbol{n}$, is constant, although it would be appreciated that both $\boldsymbol{A}$ and $\boldsymbol{n}$ increase with intensifying loads.
(2) The TRCA is linearly proportional to the applied load no matter whether the surface is deforming elastically or plastically. This is an important result and markedly different from that obtained with the array of asperities of the same height.
(3) At the loads normally used in engineering practice (usually expressed in terms of nominal pressure, i.e. load divided by the apparent geometric area of contact (ACA)), the surface roughness would only be compressed by about 10% (Thomas 1982).
(4) At loads greater than this value, the underlying bulk material reaches a yield stress and deforms plastically. The physical meaning of this argument is shown in Fig. 6.4, and it explains why the normal contact of bodies does not compress the surface asperities out of existence.

The applied load, $\boldsymbol{W}$, produces a mean contact pressure of value of $\boldsymbol{3Y}$ at the asperity contacts if they are assumed plastic deformation. The TRCA is $\boldsymbol{A}$, so that at these contacts

$$\boldsymbol{W} = 3\boldsymbol{YA} \tag{6.7}$$

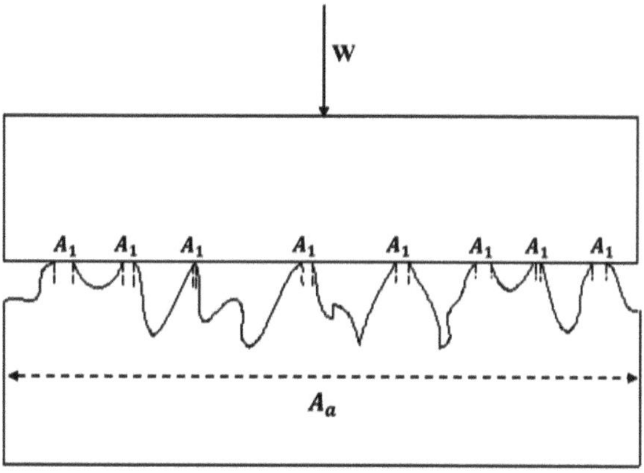

Fig. 6.4 A schematic diagram model of the contact between a smooth surface and a rough surface at the point of macroscopic plastic deformation

At the level of the bulk material, the same ***W*** would produce stress ***Y*** when this material becomes plastic deformation so that

$$W = YA_a \tag{6.8}$$

where A_a is the apparent area, i.e. the total geometric area of contact.

Most contacts of rough surfaces, however, would be between both bodies. Based on this simplified solution, theoretical assumptions for a contact model between the shoe and floor surface need to be developed and formulated.

6.2.4 Statistical Theories of Rough Surface Contact

One of the first statistical theories for the contact of rough surfaces was presented by Greenwood and Williamson (1966) and is still widely cited. Although later theories have progressively removed some of the simplifications made by Greenwood and Williamson (1966), they support the broad conclusions of this model. In the theory model of Greenwood and Williamson, it is assumed that all the contacting asperities have spherical surfaces of the same radius and that they deform elastically under load (Greenwood and Williamson 1966).

Figure 6.5 illustrates a simple, but unrealistic contact model between a nominally flat rough surface and a plane one assumed in the model of Greenwood and Williamson. The rough surface is covered with a large number of asperities, equal in shapes but with different heights.

The height of an individual asperity above the reference plane is z. If the separation between the reference plane and the flat surface, d, is less than z, then the

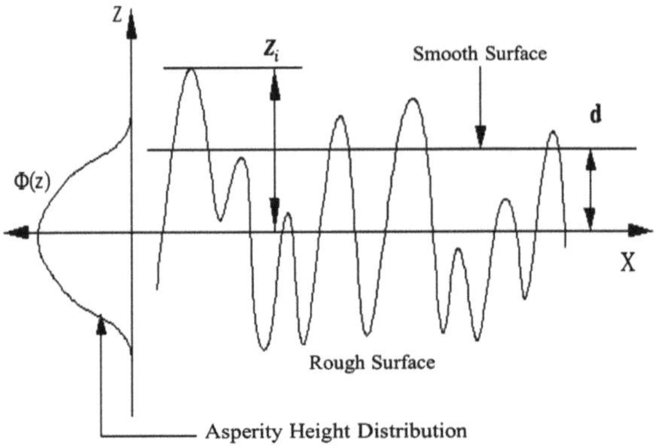

Fig. 6.5 A schematic diagram model for the contact between a rough surface and a smooth surface (Greenwood and Williamson 1966)

asperities would be elastically compressed and would support a load $\boldsymbol{W}$ that can be predicted from the Hertz's theory:

$$\boldsymbol{W} = \left(\frac{4}{3}\right)\boldsymbol{E}\boldsymbol{r}^{\frac{1}{2}}(z - \boldsymbol{d})^{\frac{3}{2}} \tag{6.9}$$

where E and r denote the composite Young's modulus and the asperity radius of curvature, respectively.

The above result shows that for this particular model, the real area of contact (RCA) is proportional to a two-thirds power of the load when the deformation is elastic. If the loads are such that the asperities are deforming plastically under the constant flow pressure, $\boldsymbol{H}$, each individual contact area, $\boldsymbol{A'}$, would be given by $\boldsymbol{2\pi\beta\delta}$. The individual load, $\boldsymbol{W_i'}$, would then be given by

$$\boldsymbol{W_i'} = \boldsymbol{HA_i'} = 2\boldsymbol{H}\pi\boldsymbol{\beta}(z - \boldsymbol{d}) \tag{6.10}$$

Thus,

$$\boldsymbol{W'} = \boldsymbol{nW_i'} = \boldsymbol{nHA_i'} = \boldsymbol{HA'} = 2\boldsymbol{HA} \tag{6.11}$$

Therefore, the RCA is proportional to the load. When sliding takes place, the RCA would be twice the value than the case of static because all the contact would be on the front face of the asperities. However, as mentioned earlier, this model is unrealistic because surface asperities of the real world solid materials have different heights following different probability distributions.

Another issue to considering is that this model is inappropriate for its oversimplified assumption. That is, the conditions of geometrical interactions between the upper smooth surface and the lower rough one were not considered. Therefore, this model should be modified to take this into account for the shoe-floor contact one.

Several contact theories between two rough surfaces were reviewed and suggested. For example, that in many practical cases of contact between metals, the majority of asperity contacts was plastic deformation. The load supported by each asperity was directly proportional to its contact area and provided that the asperities deform independently. As a result, the TRCA for the whole surface was proportional to the normal load and independent of the detailed statistical distribution of asperity heights.

For the development of a contact model between the shoe and floor surface, it can be assumed that the contact is most likely to be elastic mode. However, as the shoe heel slides over the floor surface, the contact situation would be drastically changed from the elastic mode to a plastic one. In addition, even the normal load remains constantly, the distribution of surface heights of both bodies would be modified largely since the shoe heel glides the number of asperity contacts would be grown and the mean contact area of each would be increased proportionally by massive plastic flows between them during repetitive rubbings. These assumptions would be a basis for the development of a contact model between the shoe and floor.

6.3 Some Geometrical Properties of Surface Texture

6.3.1 Introduction

Surface finishes should be considered at the design stage, with the designers specifying the footwear and flooring industry, where the surface texture required provides desired performance for the optimum slip resistance. This implies that the designer of both industries must have necessary knowledge to understand the surface topographies and analysing information on how they are measured and interpreted.

Production engineers must then plan the work so that the machine tools used are capable of producing that finish since the end products are almost entirely the result of machining processes. The measurement of surface texture is, therefore, an important step for the assessment of slip resistance properties and eventually for the prevention of pedestrian slip and fall incidents. Therefore, studies on this matter should be based on thorough comprehension for the surface texture of both pairing bodies. Detailed information on surface texture is further discussed in the following sections.

6.3.2 Surface Texture

The geometric shapes of ordinary surfaces are controlled by the characteristics of finishing processes, which they are produced. Close analysis of these surfaces shows that they are rough on a microscopic scale. Figure 6.6 shows that two major features: roughness and waviness constitute the surface texture (Dagnall 1980).

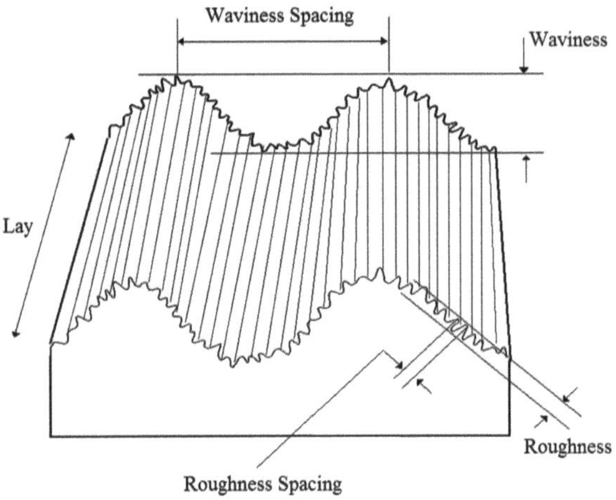

Fig. 6.6 A schematic diagram of surface texture with two major characteristics —roughness and waviness (Dagnall 1980)

The roughness and waviness are often called as primary and secondary texture, respectively. The surface texture has irregularity heights, spacing, and shapes which may look perfectly even and do not vary at all across the surface (see, Fig. 6.6). However, the surface texture may have either irregular or repetitive peak and valley patterns. Surface irregularities are classified into three categories: form errors, waviness, and roughness (Halling 1976). Based on this criterion, the three terms can be defined as follow:

- Error of form:
  - is the general shape of the surface, neglecting variations due to roughness and waviness.
  - deviates from the desired shape due to errors inherent in the manufacturing process.
- Waviness:
  - is called as macro-texture.
  - is a component of the texture upon which roughness is superimposed.
  - takes the form of relatively long wavelength variations in the surface profile.
- Roughness:
  - is called as micro-texture.
  - is the small-scale roughness of the surface associated with the inherent production process.
  - is generally interested in geometrical variations from the tribological point of view.

The above three geometrical characteristics are shown schematically in Fig. 6.7. Because materials are manufactured and finished by innumerably different processes, it would be arbitrary to distinguish amongst roughness, waviness, and form

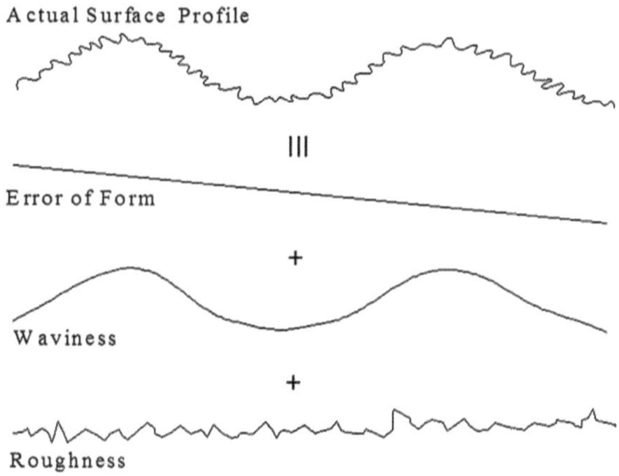

Fig. 6.7 A schematic diagram of the geometric components of a solid surface

error and have little practical significance to quote a single dividing value amongst them. Therefore, these distinctions are qualitative, yet nevertheless of considerable practical importance (Dagnall 1980).

Statistical analysis of the undulations of a surface normally shows that a very wide range of wavelengths is present, ranging from a fraction of a micrometre to many millimetres. It is likely that the range of wavelengths detected is limited only by the resolution of measuring instruments. It has been common practices to describe different wavelength bands in different terms, referring to wavelengths of the order of micrometres as roughness and the longer wavelengths as waviness. However, such a sub-division is quite illogical, and the only practical necessity is to be aware of which undulations are functionally important in any particular application.

In tribology, the individual points of contact between two solid surfaces have dimensions of the order of micrometres, and, therefore, concerned with the amplitudes of wavelengths of the same order, i.e. with those wavelengths which have been conventionally described as roughness. The peaks of such surface undulations are known as asperities and friction and wear events both arise from the contacts between such asperities on opposing surfaces.

In this sense, it seems to be reasonable to assume that both frictional and wear behaviours at the shoe-floor sliding interface seem to be depended on some ways on the stresses arising at such asperity contacts and that the stresses themselves are likely to be depended on the heights of asperities above the general level of the surface. Therefore, it could be speculated that tribological behaviours of the shoe-floor interface during repetitive contact-sliding movements would be greatly affected by the statistical distributions of asperity heights relative to the general level of both surfaces. In order to investigate such behaviour, it must be able to measure and describe these distributions.

6.4 Measurement of Surface Topography

6.4.1 Surface Texture Analysis

It has been known that a typical surface might have over 100,000 peaks (Arnell et al. 1991). The problems of measuring heights and locations of each of these peaks on each of two surfaces would be challenging. The problem of analysing the interactions between two sets of peaks when the surfaces are loaded together under normal and tangential loads would be overwhelming.

Any practical approach to this problem must, therefore, use measurements from a small but representative sample of the surface. That is, a sample chosen to be of such a size that there is a high probability that the surface lying outside the sample measured is statistically similar to that lying within it. It shall be described, therefore, the instrument used to obtain such measurements and the ways in which the measurements are interpreted to give descriptive parameters for the surfaces.

When observed on a microscopic scale, all solid surfaces are found to be uneven. In this limit, the surface irregularities would be on the scale of individual atoms and molecules. However, the surfaces of even the most highly polished engineering components show irregularities appreciably larger than atomic dimensions, and many different methods have been employed to study their surface topographies and understand topographic features.

Some involve examination of the surface by electron or light microscopes, or by other optical methods, whilst others employ the contact of a fine stylus, electrical or thermal measurements, or rely on the leakage of a fluid between the surface and an opposing plane (Arnell et al. 1991). Perhaps the highest resolution can be achieved by the techniques of scanning tunnelling microscopy (STM) or atomic force microscopy (AFM), which can resolve individual atoms; but for most engineering surfaces less sensitive methods are adequate to study their topographies. For the examination of very fine surface features, especially compliant surfaces such as elastomers and polymers, optical interferometry has clear advantages over stylus methods (Bhushan et al. 1988).

Because of the advantage of non-mechanical contact with a surface when measuring surface profile, the recently developed a laser scanning confocal microscope (LSCM) and scanning electron microscope (SEM) are used for the measurements of surface topographies and their specific characteristics in this study. Detailed features on the LSCM instrument are demonstrated in Sect. 6.4.3.

6.4.2 Surface Profilometry

One of the most common methods of assessing the surface topography is a stylus profilometer, which is illustrated in Fig. 6.8.

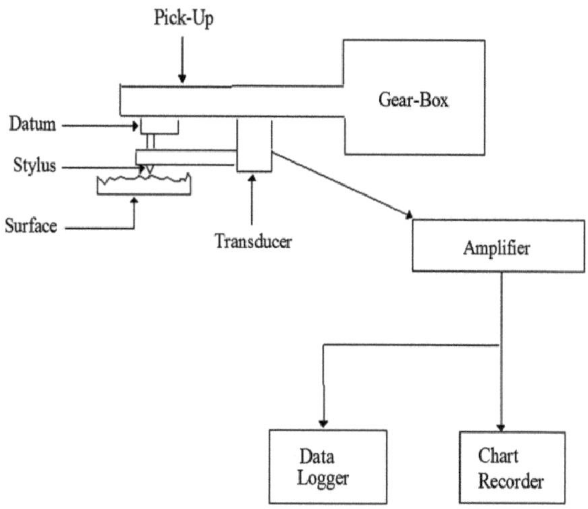

Fig. 6.8 Schematic views of a simple stylus profilometer and the operation principles. The stylus moves steadily over the surface under examination, and its vertical displacement is recorded on a moving chart or digitised for computer processing (Thomas 1982)

In the profilometer, a very fine diamond stylus is dragged smoothly and steadily across the surface under examination. As the stylus travels over the surface, it rises and falls. Its vertical displacement is converted by a transducer into an electrical signal that is amplified and, in the simplest form of the instrument, moves the pen of a chart recorder. The graph drawn by the pen represents the vertical displacement of the stylus as a function of the distance travelled along the surface.

The graphical representation of the surface profile generated by a stylus profilometer differs from the shape of a genuine cross-section through the surface for several reasons. The major difference is due to the different magnifications employed by the instrument in the horizontal and vertical directions.

The vertical extent of surface irregularities is nearly always much less than their horizontal scale. It is, therefore, convenient to compress the graphical record of the surface profile by using a magnification in the vertical direction, which is greater than that in the plane of the surface.

The ratio of the magnifications depends on roughness profile of the surface. In Fig. 6.9, the shape of a real surface is compared with profilometer recordings of the same surface at different magnification ratios (vertical vs. horizontal) of 5:1 and 50:1.

Although the amplitudes and wavelengths of the irregularities are carefully recorded on both graphs, the surface slopes appear much steeper on the profilometer records than they really are. The distorting effect of this horizontal compression must be recognised in interpreting them.

As clearly shown in Table 4.1 from Sect. 4.3.2, unavoidable limitations seem to be faced by using this device although it is performed well for the general measurements of surface texture. Followings are the summary of possible restrictions from the stylus type profilometer to perform fine measurements of surface roughness for both shoes and floors:

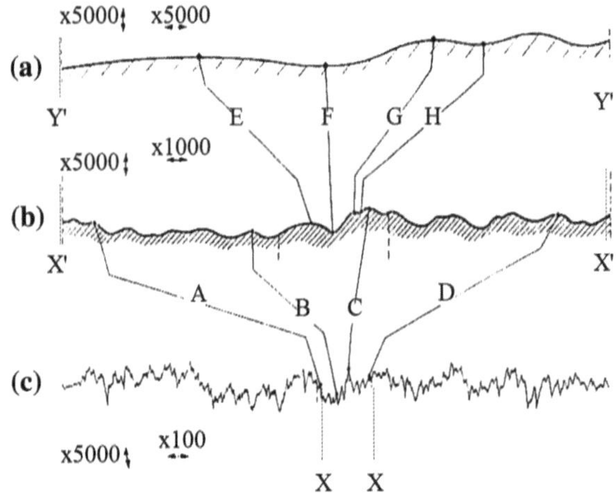

Fig. 6.9 A schematic diagram illustrating how the profile shape varies as the horizontal magnification is reduced (Dagnall 1980). **a** The profile of a real surface, magnified ×5000 equally in all direction; **b** the same surface with a ratio of 5:1 between vertical and horizontal magnifications; **c** as for (**b**), but with a ratio of 50:1 between vertical and horizontal magnifications

(1) Styli of the profilometer are made of diamond. The shapes can change from one manufacturer to another. For example, chisel-point styli with tips (0.25 μm × 2.5 μm) may be used for detection of bumps or other special applications (Bhushan 2000). Conical tips are almost exclusively used for microroughness measurements (Bhushan 2000). According to the international standard (ISO 3274 1996), a stylus is a cone of a 60°–90° included angle and a spherical tip radius of curvature of 2, 5, or 10 μm. The radius of a stylus ranges typically from 0.1–0.2 μm to 25 μm with the included angle ranging from 60° to 80°. The portion of the stylus tip that is in contact with the sample surface, along with the known radius of curvature, determines the actual radius of the tip with regard to the feature size. The stylus cone angle is determined from the cleave and grind of the diamond chip and is checked optically or against a standard (Bhushan 2000). The stylus load ranges typically from 0.05 to 100 mg. Long-wave cutoff wavelengths usually range from 4.5 μm to 25 mm. Short-wave cutoff wavelengths range typically from 0.25 μm to several millimetres. The scan speed ranges typically from 1 μm/s to 25 mm/s. The sampling rate ranges typically from 50 Hz to 1 kHz.
However, the combination of finite tip radius and included angle prevent the stylus from penetrating fully into deep narrow features of the surface. Special styluses with chisel edges and minimum tip radii as small as 0.1 μm can be used to examine fine surface details where a conventional stylus would be too blunt. But, all the stylus methods inevitably produce some "smoothing" errors of the surface profile due to the finite dimensions of the stylus tip. In principle, these problems have been known ever since the contact stylus instruments have existed.

(2) Since systematic errors appearing at the beginning of the measuring chains between the stylus tip and the surface are transmitted through the whole measuring sequences, their detailed investigation is of fundamental importance (Hillmann et al. 1984). In some applications, this can lead to significant errors.

A finite size of stylus tip distorts a surface profile to some degree (Radhakrishnan 1970; McCool 1984). This means that the radius of curvature of a peak may be exaggerated, and the valley may be represented as a cusp. A surface profile containing many peaks and valleys of radius of curvature of about 1 μm or less or many slopes steeper than 45° would probably be more or less misrepresented by a stylus instrument.

Another error source is due to stylus kinematics (McCool 1984). A stylus of finite mass held in contact with a surface by a preloaded spring may, if traversing the surface at a high enough velocity, fail to maintain contact with the surface being traced. Where and whether this occurs depends on the local surface geometry, the spring constant to the mass ratio, and the tracing speed. It becomes clear that a trace for which stylus contact has not been maintained presents inaccurate information about the surface microroughness.

(3) Stylus load also introduces error (Bhushan 2000). The load on the stylus, although small, may nevertheless distort or damage the surface so that non-mechanical contact method such as an optical interferometry is recommended for measuring the surface profile (Bhushan et al. 1988). A sharp stylus even under low loads results in an area of contact so small that the local pressure may be sufficiently high to cause significant local elastic deformation of the surface being measured. In some cases, the local pressure may exceed the hardness of the material, and plastic deformation of the surface may result (Bhushan 2000).

Styli generally make a visible scratch on softer surfaces, for example, some steels, silver, gold, lead, and elastomers (Poon and Bhushan 1995a, b, 1996). The existence of scratches results in measurement errors and unacceptable damage. Hence, it is important to select stylus loads low enough to minimise plastic deformation.

(4) Another serious disadvantage is that the stylus tip has to be in direct contact with the surface, which may change the surface and/or stylus and cause contamination. Moreover, due to the mechanical interaction, the scan speeds are significantly slower than with optical methods. Because of the stylus angle, stylus profilometers cannot measure up to the edge part of a rising structure, causing a "shadow" or undefined area, usually much larger than what is typical for optical systems.

(5) Most of the profilometers show severe limitations in terms of the surface features detectable and difficulties arisen when three-dimensional data sets of surfaces are required. Three-dimensional approaches may provide further detailed information on micro aspects of the surface geometry. This information seems to providing a vital source for the investigation of wear behaviours of the shoe and floor surfaces and the understanding of wear mechanisms at the sliding interfaces between them. At present, it is scarce to find detailed information on sliding friction induced wear developments of the shoe sole/heel and floor surfaces and their shared effects on slip resistance performance between the shoe and floor surface.

(6) Frictional and wear behaviours of the sliding interface between the footwear and underfoot surfaces show constant changes with certain fixed patterns and/or random ones during repetitive sliding friction measurements (Kim 2015a, b, c, d, 2016a, b). As a result, friction caused wear developments at the shoe-floor sliding interface can generate different tribo-physical situations depending upon wear progresses (Kim 2015a, b, c, d, 2016a, b).
This means that formation and travelling of wear materials between the floor surface and shoe heel seem to take place in the following three conceivable ways (Kim 2015a, b, c, d, 2016a, b):

- Straight move: wear flow from a shoe heel to a floor surface or a floor surfaces to a heel surface;
- Reverse move: wear flow from a shoe heel to a floor surface and a floor surface to a heel surface; and
- Joint move: repeated wear flows from a heel surface to a floor surface to a heel surface to a floor surface.

As discussed in the introduction, different wear mechanisms and flows should be fully understood by investigating fundamental aspects of their involved tribo-physical characteristics as a first step. A systematic observation for the geometric features of both bodies during recurring sliding processes is, therefore, a critical point to step forward to understand their principles on fictional and wear behaviours of the shoe heels and floor surfaces.

(7) A stylus of the profilometer could be bounced on a polymeric material such as shoes due to the high resilience of the material properties (Chang et al. 2003). This also could affect the result of accuracy from the profile readings of surface roughness.

(8) The profilometer may also cause to the inaccuracies in measuring surface roughness when there are transferred wear products such as polymer films, flooring materials, dirt, and grits and deep scratched worn areas.

In spite of the above-reviewed limitations, the profilometer remains as an outstanding instrument for studying surface texture and evaluating its descriptive parameters. During the initial stage of experimental works in this book, the Talysurf 5 surface roughness measurement system, which had a spherical tip of 12 μm radius was used for the measurements of surface roughness of some floor samples.

As mentioned above, however, it was urged to find a more advanced way to investigate the surface profiles of shoe and floor specimens because of the limitations in detectabilities of fine details on surface features due to a rather large size of spherical tip of the Talysurf 5 instrument. In the later part of this book, one of the most recently developed instruments, which had clear advantages over the stylus method, was adopted to measure the surface topographies for shoe and floor samples before and after dynamic friction measurements.

Following section describes the principles of confocal imaging and operating details of a laser scanning confocal microscope used in this book.

6.4.3 Laser Scanning Confocal Microscope

6.4.3.1 Introduction

Conventional methods for measuring surface profiles usually have involved profilometers as mentioned in the previous chapter. Profilometers, however, show severe limitations in their measurements such as fine details of surface features and difficulties in detecting three-dimensional data sets of surfaces. Alternative methods that have been explored are stereo microscopy, reflected light interference microscopy (RLIM) and scanning electron microscopy (SEM). But, these methods have proven to be severely limited either by the depth of field that can be obtained, difficulties associated with obtaining and interpreting images or the prohibitive costs involved (Hutchings 1992; Anamalay et al. 1995).

A new type of light microscope, a laser scanning confocal microscope (LSCM), is a fresh development in the field of microscopy that has the potential to revolutionise surface analysis. This instrument is similar to conventional confocal microscopes and can achieve similar resolutions. However, one of the distinct advantages from this new microscope is its ability to generate rapidly nondestructive optical sections in thick opaque specimens such as shoes and floor surfaces. Another advantage of this instrument is that samples need not be specially prepared or mounted and as long as they are able to sit on the stage of a conventional microscope, their surface topographies can be captured.

6.4.3.2 The Principles of Confocal Imaging

The LSCM differs from a conventional optical microscope in that it illuminates and images the sample one point at a time. The image is built up, pixel by pixel, by scanning either a laser beam or the sample. Resolution is determined by the diffraction-limited laser spot size. The LSCM has better lateral resolution than the conventional optical microscope (Wilson and Sheppard 1984; Fellers and Davidson 2007). Digital image processing can be used to improve the image signal-to-noise (S/N) ratio and contrast.

The term "confocal" indicates that both condenser and objective lenses are focused on the same point in the object. In the epi-illuminating mode, a single objective lens serves as both condenser and objective for the epifluorescent and reflected light (see, Fig. 6.10). A pinhole in front of the photodetector obstructs most of the scattered, reflected or fluorescent light from out-of-focus planes. A thin layer of a thick opaque object can be imaged clearly; the pinhole ensures that image information is obtained from only one particular plane in the specimen. The pinhole also reduces image blurring due to light scattering, thus increasing the effective resolution (Inoue 1990; Sanderson et al. 2014).

Optical sectioning makes it possible to reconstruct, by the digital image processing software which combines two different types of images from a three-dimensional

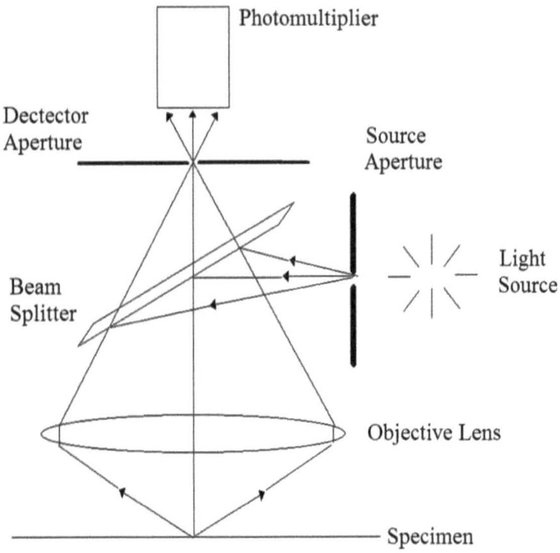

Fig. 6.10 Schematic descriptions for depth discrimination in the confocal scanning microscope

image data block. The data block is comprised of a stack of two-dimensional images acquired over the depth of a thick opaque object. The image slices in such a three-dimensional block can be added to give a two-dimensional image that is in focus throughout the entire depth of the original block—an image which cannot easily be generated with a conventional optical microscope.

Like other confocal microscopes, the LSCM takes successive ‘slices’ or optical sections of an image, with each optical section being of a very narrow focal plane. Only the regions of the sample that is in focus are picked up by the detector, hence discarding all out of focus regions. For instance, as shown in Fig. 6.11, only regions *A* to *D* that is in focus are picked up in that particular scan. Thus, by starting at the bottom of a specimen and taking successive image scans (optical sectioning) until the top of the specimen is encountered, it is possible to capture the entire 3D surface topography of the sample.

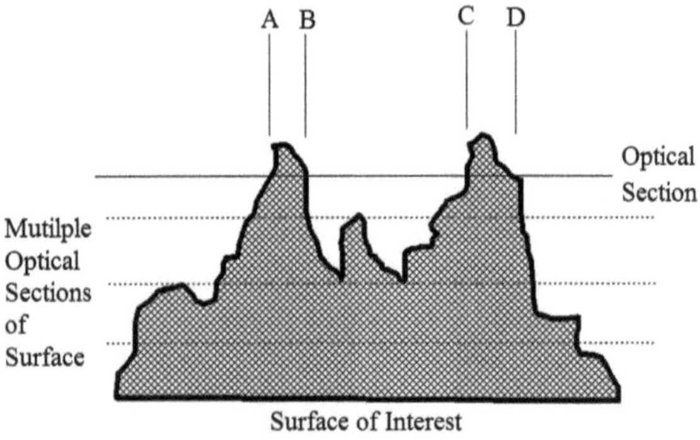

Fig. 6.11 Schematic descriptions for multiple optical sections to capture surface topography

From the "stack" of successive optical sections of the surface so obtained, it is possible to extract the surface topography. Computer software was specially developed for the study in this book to enable a more objective analysis of surface texture. The software allows the compilation of a "maximum brightness projection" and a "height encoded image", which facilitate surface analysis.

The maximum brightness projection is an image that is formed by the points of maximum brightness for each pixel location and for each image in the stack of optical sections. This results in a reflectance image of the surface which is in focus at all locations with sharp definition of surface features. The height encoded image is an image whose pixel intensities enumerate the location in the stack of optical sections they are taken from. It is analogous to a contour diagram of the surface, with each shade of grey in the image representing a particular height.

By analysing the height encoded image, basic but important parameters for surface roughness can be obtained. The developed software allows to calculate conventional surface roughness parameters such as the centre line average (***Ra***), root measure square roughness (***Rq***), maximum peak-to-valley height (***Rt***), maximum mean peak-to-valley height (*Rtm*), maximum departure of the profile above and below the mean line referred to as ***Rp*** and ***Rv***, and mean of maximum departures of the profile above and below the mean line, referred to as ***Rpm*** and ***Rvm***, respectively. Detailed information on the surface roughness parameters is found in Sect. 6.8.2. The collected information for the surface roughness parameters can be saved into a computer for the further analysis.

To measure the surface characteristics, the first step is to obtain successive optical sections of the surface to be studied. The stack of optical sections is then processed by the software installed and a height encoded image of the surface is produced. The height encoded image is then used subsequently by the software for the calculations of basic surface roughness parameters since it contains all the important surface information extracted out of the stack of optical sections. Thus, instead of working with multiple optical sections, a single image containing the desired data is dealt with. This reduces processing time and computer memory requirements.

The operator selects sections of the height encoded image that is to be analysed and draws a line over the region of interest (here the line length denotes the "assessment length") or encloses the region of interest with a rectangle. In the case of a line of interest, the following phases apply. The line specifying the assessment length, to keep with the convention of typical profilometers, is split up into a number of "sampling lengths", each of equal length.

The software allows the number of sampling lengths per assessment length to be varied, to suit the operator's requirements. The software then calculates certain roughness parameters using algorithms that have been programmed for the surface studies of shoe and floor samples in this book. The surface roughness parameters measured and calculated from this microscope are saved in the computer for the further analysis.

6.4.3.3 Laser Scanning Confocal Microscope (LSCM)

A Bio-Rad MRC-600 LSCM with improved MRC-600 optics, attached to a Zeiss Axiophot microscope, was used for the studies of surface texture and the measurements of surface roughness of both shoe and floor specimens in this book. Figure 6.12 shows the total system. The light source was a 25 mW argon ion laser with excitation wavelengths at 488 and 514 nm (Bio-Rad Microscience, 1988). In all cases, a neutral density filter was used to reduce the intensity of the laser.

The automatic gain control for the photomultiplier was disabled to minimise the non-linearity in the response which was a problem in quantitative measurements. All observations were made with three different Plan-Apochromatic objective lens magnifications—$\times 20$, $\times 50$, and $\times 100$—with a numerical aperture of 1.5. The resolution of the system was asymmetric.

To generate a cross-sectional image, each specimen of the shoe heels and floor surfaces was firstly oriented perpendicular to the horizontal scanning direction. In the LSCM, as in a conventional light microscope, the focal plane is fixed relative to the optics. In order to scan the image plane through the depth of an object, the object itself must be moved along the optical axis.

The cross-sectional (x-z) images presented for this study were constructed from a series of horizontal line scans acquired whilst stepping the sample stage vertically in the z-direction (see, Fig. 6.13). For the step size, 1 μm was used initially. According to the geometry structures of the shoe and floor specimens, 20 up to 32 scans per step were performed and the S/N ratio was optimally adjusted to each surface roughness.

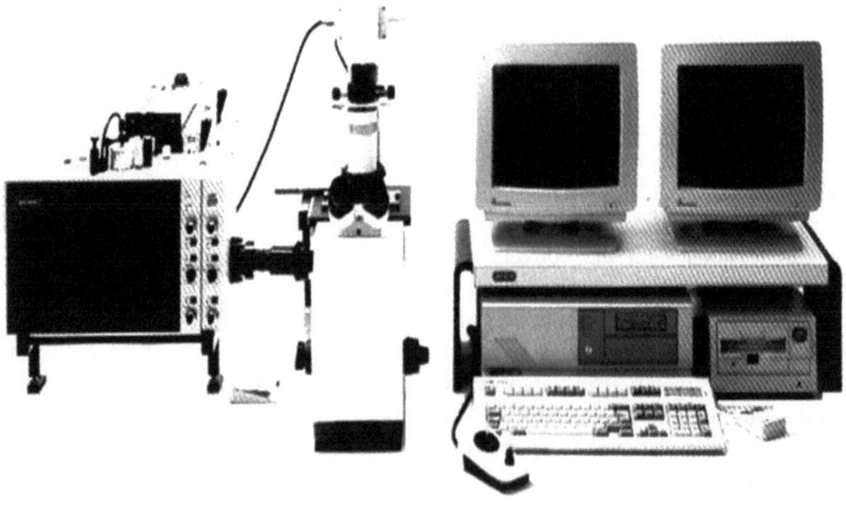

Fig. 6.12 A photographic image of the Bio-Rad Lasersharp MRC-600 LSCM Model

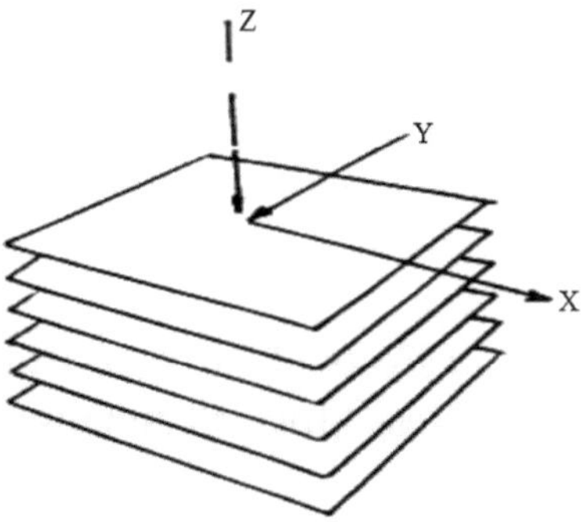

Fig. 6.13 Schematic illustration of representation of Z-series of images

6.5 Importance of Surface Analysis for Slip Resistance Assessment

The recent literature reports that surface roughness of the shoes and floors have substantial effects on slip-resistance performance under a range of walking environments (Grönqvist et al. 1990; Grönqvist 1995; Rowland et al. 1996; Wilson 1996; Kim 1996a, b, c, d, 2004a, b, 2005a, b, 2006a, b, 2015a, b, c, d, 2016a, b; Manning et al. 1998; Kim and Smith 1998a, b, 1999, 2000, 2001a, b, c, 2003; Kim et al. 2001, 2013; Kim and Nagata 2008a, b). Surface roughness provides necessary drainage spaces to avoid squeeze film formations under contaminated environments. Micro- and macro-tread patterns on the shoe heel/sole area increase slip-resistance properties by providing spaces for the removal of pollutants and lead to an increase in direct contacts with the floor surface. Thus, large-scale roughness and/or tread patterns are commonly designed into the shoe bottom regions (Kim and Nagata 2008a, b; Kim et al. 2013; Kim 2015a, b, c, d, 2016a, b).

Geometric characteristics of the floor surface, however, could rather drastically improve slip resistance properties than the shoe ones. Because floor surfaces may provide sharper, taller, and tougher asperities that would be enough to extend upward through lubricating films sufficiently to engage with the bottom areas of heel surfaces in a manner like sandpapers (Kim and Nagata 2008a, b; Kim et al. 2013; Kim 2015a, b, c, d, 2016a, b). Hence, the surface roughness of the shoes and floors should be fully investigated with friction and wear behaviours when analysing slip-resistance properties.

Although key concepts and theory models to predict slip resistance properties from known surface characteristics have not developed, if the surface features of shoes and floors and their interactions could be quantitatively measured and systematically analysed, then our understanding for the multifaceted characteristics of friction and wear behaviours from shoes and floors would be considerably enhanced. On a broader scale, this may also assist to improve design aspects of the footwear and floor surfaces that consequently lead to substantial reductions in the slip and fall incidents (Kim et al. 2013; Kim 2015a, b, c, d). Therefore, it would be necessary to measure surface roughness of the shoe and floor and to analyse their effects on slip resistance performance.

6.6 Effects of Surface Roughness on Slip Resistance Performance

Almost all surfaces are rough on a microscopic scale and comprise an aggregation of micro- and macro-asperities. That is, most of the solid surfaces have surface roughness and the variations in the surface roughness profile can be represented by a random arrangement of peaks and valleys. When two such surfaces are in contact, they touch only at tiny discrete areas where their highest asperities are in contact (Bowden and Tabor 1954; Tabor 1974; Kragelsky and Alisin 1981; Kim and Nagata 2008a, b; Kim et al. 2013; Kim 2015a, b, c, d). The local pressure at the contact regions is then high enough to cause plastic deformation of the asperities even at the lightest load.

If at least one of two sliding surfaces is of a viscoelastic material such as the shoe-floor sliding system, the variation of COF with the normal pressure could have practical consequences. This means that a contact-sliding system between the shoe heel and floor surface would be an elastoplastic state and have an interlocking mechanism. The interlocking mechanism would be governed by a number of factors such as asperity shape, size and distributions of the shoe heels and floor surface, surface properties, normal load, surface conditions (dry and/or lubricated), and sliding speeds under which the contacts occur.

During repeated rubbings, topographic features of both pairing surfaces would be continuously changed by friction and wear processes. That is, as the shoe heel frequently slides over the floor surface, variations in the surface topographies of two bodies may occur at the same place and/or at different locations according to their surface profile structures.

Recently, Kim (2015a, b, c, d) examined the progressive wear and surface changes of three shoe surfaces during slip resistance measurements. This study showed that variations in the surface texture of the shoe heels had a major effect on the slip resistance properties. Topography changes of floor surfaces were also analysed before and after slip resistance measurements (Kim and Smith 2000; Kim 2004b; Kim et al. 2013). Those studies found that embedding or transferring of polymeric materials from shoe heels into the valley areas of floor surfaces was a major cause for the changes of floor surface roughness.

Other findings also showed that surface roughness parameters were well correlated with the standard deviation of peak heights and the changes in the surface roughness within the contact areas, as well as the comparative engagements of two surfaces between the shoe heels and floor specimens during dynamic friction tests (Kim and Smith 1998a, b, 1999, 2000; Kim 2004a, b, 2015a, b, c, d, 2016a, b; Kim and Nagata 2008a, b). All those studies clearly identified that surface topographies of the shoes and floors underwent notable alterations during the slip resistance measurements. As a result, surface changes of the shoes and floors throughout the course of repeated friction and wear developments had a major effect on the slip resistance performance.

The recent studies also found that wear developments and wear-induced surface variations were more severe than expected and occurred at a very early stage of sliding friction. That is, slip resistance properties between the shoes and floors seem to largely depend not only on the friction develop when a slip starts, but also on how the friction change as a slip progresses. Therefore, it becomes clear that surface topographies of both shoes and floors should be routinely monitored and carefully analysed with the measurements of slip resistance properties.

6.7 Quantifying Surface Roughness

6.7.1 Introduction

If every different surface is given a unique number, then it would be helpful to convert roughness assessment from a subjective into an objective procedure. In another word, it would be able to measure surface texture. However, a surface texture is very complex, and no single parameter is found to quantify all its various topographic characteristics.

Fortunately, it is known that only a few of these surface characteristics have practical importance or can be reliably related to performance in most applications (Hutchings 1992). So, if these few surface characteristics can be quantified, it shall be able to state that it can measure surface texture for practical purposes.

The following sections explore further ways in which surface finishes can be expressed numerically, and the conditions for quantifying them. The main criterion for selecting a surface texture parameter would be:

> Can it be related to performance or to the production processes it is controlling?

In other words,

> Does the roughness parameter value change rapidly or slowly, or regularly or irregularly with the function it is monitoring?

With the recognition of the main purpose of selecting surface roughness parameters, descriptions are made to represent the surface roughness parameters by utilising their statistical properties, as outlined in the following sections.

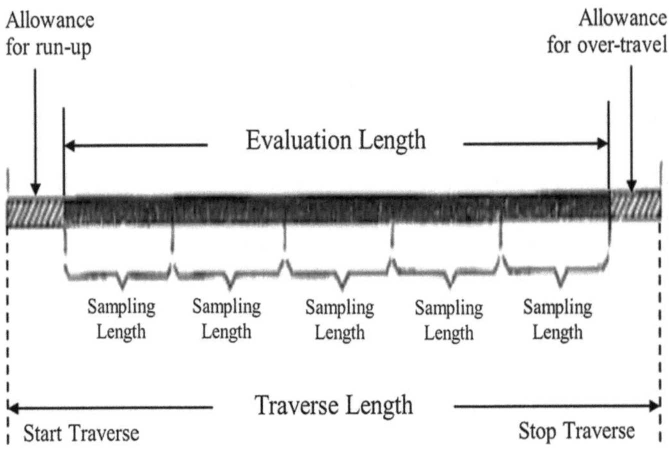

Fig. 6.14 Schematic demonstration on a relationship amongst sampling, evaluation and traverse lengths

6.7.2 *Measuring Lengths*

Figure 6.14 shows that there are three characteristic lengths associated with the numerical assessment of surface texture: sampling, evaluation, and traverse lengths, respectively (Dagnall 1980).

A brief definition for the three lengths are summarized in the below:

(1) The sampling length is a length of the surface which a single assessment of a parameter is made, being the length over which the parameter to be measured has statistical significance without being long enough to include irrelevant details. The number of sampling lengths in the evaluation can vary.
(2) The evaluation (or assessment) length is a length of a surface over which a measurement is made. This length may include several sampling lengths and the measurement then being the average of more than one assessment of the parameter.
(3) The traverse length is the total length of surface traversed by the stylus in making a measurement. It is normally greater than the evaluation length due to the necessity of allowing a short over travel at either end to ensure that mechanical and electrical transients are excluded from the measurement.

6.7.3 *Reference Line*

A prime requirement for quantifying surface roughness is the provision of some datum within the roughness profile to which measurements can be related. In the surface metrology, it is not measuring dimensions of the bulk material, but deviations from an ideal shape (Schaffer 1988). It follows that it must use this ideal shape (or as close an approximation to it as is practical) as the reference line.

For practical purposes, the reference line adopted for most roughness parameters is the centre line of the profile (sometimes called as a "mean line"). This is a straight line that runs centrally through the peaks and valleys along the line that would be left if the peaks had been levelled to fill the valleys.

Mathematically, it is positioned such that, within the sampling length, the sum of the areas enclosed by the profile above the centre line equals the sum of those below it. Within each sampling length, the line is theoretically straight although its slope relative to the nominal shape of the roughness profile depends on waviness. So, when the centre line is determined graphically from a recorded roughness profile, it can be drawn as a straight line.

However, when it is determined by instrumental means from the electrical waveform representing the roughness profile, the centre line may be curved, but this would not significantly affect the measured results (Dagnall 1980).

6.7.4 Traditional Surface Roughness Parameters

Accurate assessments for the topographic characteristics of the surface texture have become increasingly important. On the other hand, surface texture is a key factor affecting the function and reliability of a component. Its measurement and analysis can prove to be an excellent diagnostic tool for monitoring the process that is produced the surface in question (Abouelatta et al. 2006).

Surface texture refers to the locally limited deviations of a surface from the ideally smooth one which is an intended geometry of the part, deviations from the nominal surface that is closer together than geometric irregularities (Schaffer 1988; Kang et al. 2008). The deviations can be categorised on the basis of their general patterns. Surface texture includes closely spaced random roughness irregularities and more widely spaced repetitive waviness irregularities.

Roughness, which is the finer random irregularities of surface texture, usually results from the inherent action of the production process instead of from the machine. These include traverse feed marks and other closely spaced irregularities produced by cutting tools, grits, or other process-related actions (Schaffer 1988).

Waviness, which is the wider-spaced repetitive deviations, can usually be attributed to the characteristics of an individual machine or to external environmental factors. It may result from such factors as machine or work deflection, vibration, chatter, heat treatment, or warping strains (Schaffer 1988). Because both process and machine-induced irregularities occur simultaneously, roughness appears superimposed on waviness.

In reality, however, surface texture cannot be measured directly. It is not possible to assign a unique value to every different surface (Schaffer 1988). However, it is possible to measure some of the inherent characteristics or parameters of surface texture.

In fact, any parameter set cannot include all parameters which can describe surface topography completely. Only some major topographic properties related to geometry, statistics, and functions can be and are necessary to be described (Dong et al. 1994).

Because one of the main concerns of this book is not a development of the definition of the surface parameter and this matter is also beyond this book, the selection of surface parameters used for this book is totally based on the following criterion.

(1) Measurement of the surface topography should be objective and the surface roughness parameters describing it also should be objective and independent of the instrument used.
(2) The surface roughness parameters defined should not be correlated.
(3) The surface roughness parameters included in a parameter set should be able to describe some major and fundamental properties of the surface topographies of both shoes and floors.

The surface roughness parameters developed over the years fall into the following three basic categories.

(1) Amplitude parameters:

These are determined solely by peak heights or valley depths, or both, of profile deviations, irrespective of their spacing along the surface.

(2) Spacing parameters:

These are determined solely by the spacing of profile deviations along the surface.

(3) Hybrid parameters:

These are determined by amplitude and spacing in combination.

Surface roughness parameters used for this book are selected on the basis of the above categories. Following sections fully describe the surface roughness parameters with their related equations and principles.

6.8 Statistical Analysis of Surface Finishes

6.8.1 Background

The irregularities of machined surfaces consist of high and low marks created by a tool bit or a grinding wheel. These peaks and valleys can be measured and used to define the conditions and sometimes the performance of the surface. There are more than 100 ways to measure a surface and analyse the results, expressed as parameters, but for most cases, only a few are specified (Nugent 2008).

The most common parameter is R_a, or Arithmetic Average Roughness. It basically reflects the average height of roughness component irregularities from a mean line. The R_a parameter provides a simple value for accept/reject decisions. It is a default parameter on a drawing if not otherwise specified, and is available even in the least sophisticated instruments. However, the R_a parameter is not a good discriminator for different types of surfaces as it is incapable of differentiating between "spiky" and "scratched" surfaces. Additional roughness parameters should be specified for this purpose, such as R_p (maximum peak height), R_v (maximum valley depth) and R_t (maximum peak-to-valley roughness height).

Each of the above roughness parameters has its own advantages and limitations. Often one parameter is incapable of defining a surface adequately. Therefore, a complete definition of a surface often involves two or more roughness parameters, and in some cases, the relationship or ratio of one parameter to another.

Detailed information on the surface finish measurement procedures, general terminology, definitions of most parameters and filtering information can be found in American Standard ASME B46.1-2002, Surface Texture, and in International Standards, ISO 4287 and ISO 4288.

6.8.2 Statistical Analysis of Surface Roughness

Surface profiles often reveal both periodic and random components in their geometric variations and such components are not revealed by R_a and R_q values. Until the recent years, the analogue output of a profilometer was usually analysed within the instrument to give a graphical record of the profile and a single-parameter description of the surface roughness, which was displayed on an analogue meter (Arnell et al. 1991). However, it has become common to digitise the surface profile, to give a record which is more amenable to statistical analysis, and to separate its periodic and random elements (Moalic et al. 1987).

In this technique, the analogue output from the amplifier is digitally sampled and recorded as a dataset of N points acquired at discrete intervals to give a sequence of height readings relative to the profile centre line (see, Fig. 6.15). The sequence can then be analysed by an inbuilt microcomputer, to extract any chosen statistical information. The two most common types of information to be extracted are the height distributions of surface texture and the amplitudes and wavelengths of periodic variations along the surface. These are further explored in the following sections.

6.8.3 Height Distribution of Surface Texture

Two-dimensional surface profiles can be considered as a height Y which is a function of distance X from a reference point on the surface. For most modern

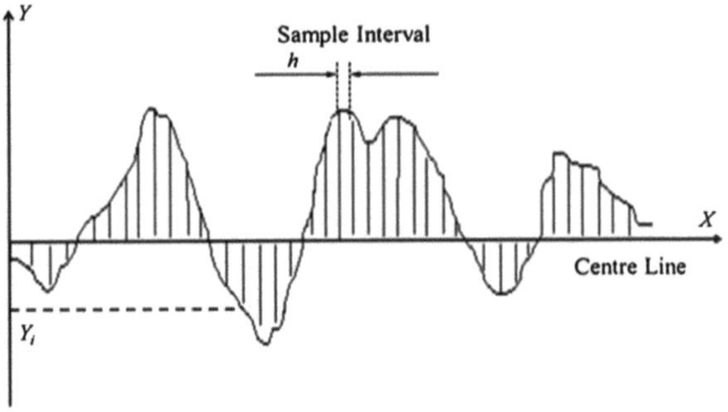

Fig. 6.15 Schematic description for typical surface height readings taken at discrete intervals

surface analysis methods, the surface profile is digitally sampled and recorded as a dataset of N points acquired at discrete intervals as shown in Fig. 6.15.

The surface is thus to be characterised using the data set Y_i (i = 1, 2, …, N) at some discrete interval h and any analysis performed on this data set would be subject to errors dependent on sample size N and sample interval h (Moalic et al. 1987).

6.8.3.1 CLA and RMS Roughness

Initially, a surface is described by its centre line average, R_a, and its root mean square, R_q, defined in terms of digitised height readings. The profile ordinate heights are treated as positive or negative with respect to a datum.

The data are later interpreted with respect to an arbitrary datum so that all ordinate (Y_i) could be treated as positive numbers. A mean line is fitted to the data by the least squares method and the surface ordinates Y_i' with respect to this line are obtained, as shown in Fig. 6.16.

The average roughness (symbol: R_a, c.l.a. for centre line average or AA for the arithmetic average) is defined as the arithmetic mean deviation of the surface height from the mean line through the profile. Whilst the root mean square (symbol: R_q, RMS) is defined as the square root of the mean of these deviations.

One of the main criticisms of the R_a roughness parameter is that it cannot distinguish between profiles of a different shape. Figure 6.17 shows an example for two very different surfaces with the same R_a roughness parameter. Despite this drawback, the R_a roughness parameter is in almost universal use for quality control (Sherrington and Smith 1987). The R_a has traditionally been the most commonly used roughness parameter in the UK, Germany, and France whilst the R_q has been the most common in use in the USA (Thomas 1981).

On the other hand, one advantage that the R_q roughness parameter offers over the R_a roughness one is that it may reflect changes in the shape of the profile when the material is conserved when the R_a does not (Leaver et al. 1974). This arises as a result of the form of the definition of R_a.

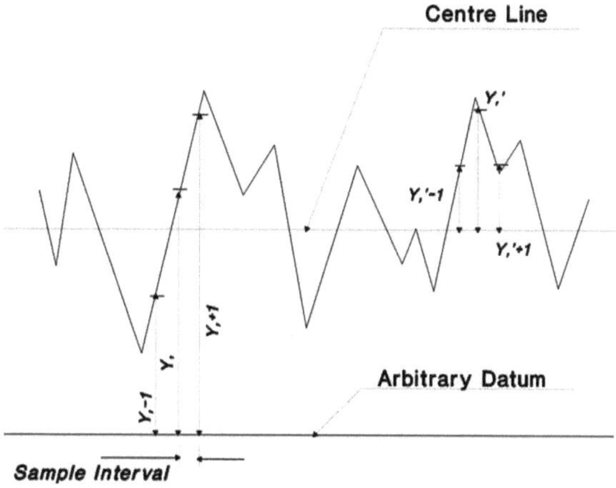

Fig. 6.16 Two-dimensional representation of a surface profile

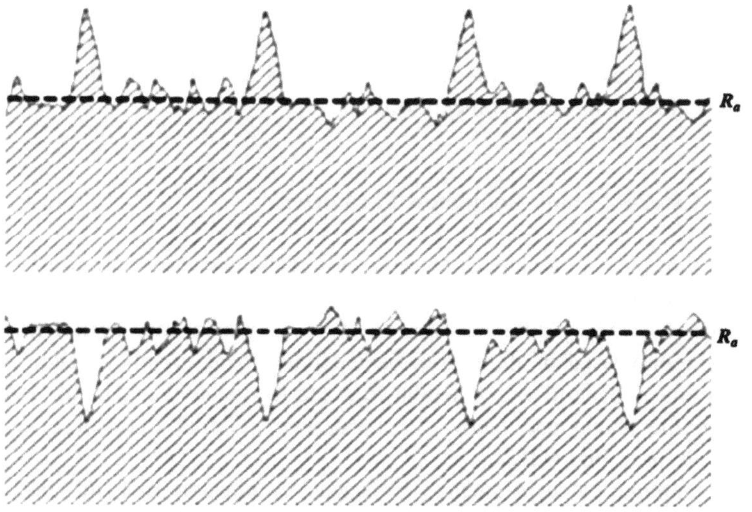

Fig. 6.17 A schematic example for two very different surfaces with the same R_a roughness

The R_a roughness parameter is assessed by reference to a mean line positioned so that the areas enclosed by the profile above and below it are equal. Any redistribution of material from one side of the mean line to the other would cause a shift in the position of the mean line and may leave the R_a value unchanged.

The R_q value, however, would be modified by the operation. This inability to identify changes in the profile shape when the material is conserved means that the R_a is less suitable than the R_q for monitoring certain surface processes (Thomas 1982).

However, as briefly described in the above, both R_a and R_q roughness parameters do not provide any information about the shapes, slopes, and sizes of the asperities or about the frequencies of their occurrence. It is, therefore, possible for surfaces of widely different roughness profiles to give the same R_a and R_q values (Sherrington and Smith 1987).

Despite this drawback, these single-parameter descriptions are mainly useful for comparing surfaces, which have been produced to different standards, but by similar methods (Arnell et al. 1991). In practice, the R_q roughness parameter is more statistically significant because it represents the standard deviation of profile heights whilst the R_a roughness one has the advantage of more common usage (Stout and Davis 1984).

For many surfaces, the values of R_a and R_q roughness parameters are similar with a Gaussian distribution of the roughness profile from the mean line:

$$R_a = \sqrt{\frac{2}{\pi}} \times R_q \cong 0.8 \times R_q \tag{6.12}$$

According to Thomas, this relationship is a good approximation for most surface types provided their height distribution is almost symmetric, even if it is not Gaussian (Thomas 1982). This relationship arises as a result of the form of the definition of R_a parameter as mentioned above.

6.8.3.2 Extreme Value Roughness Parameters

It is useful to have a measure of the extremes of the departure of a profile such as the maximum roughness height rather than the mean height that the R_a roughness parameter gives. Many different forms of these parameters have been defined. The most commonly used of these are; the maximum peak-to-valley height, R_t, the maximum mean peak-to-valley height, R_{tm}, and the mean of maximum departures of the profile above and below the mean line, referred to as R_{pm} and R_{vm}, respectively (see, Fig. 6.18).

These extreme value parameters are sensitive indicators of high peaks and/or deep scratches in a surface (Sherrington and Smith 1987). The maximum peak-to-valley surface parameter (R_t) has been used to assessing the roughness of both floor surfaces and shoes (Harris and Shaw 1988; Proctor 1993; Manning and Jones 1994; Kim et al. 2013; Kim 2015a, c, 2016a, b). However, they simply used this parameter so that it should be noticed that there were some relationships amongst these extreme roughness parameters.

The vertical distance between the highest and lowest points of the profile within the sampling length has been given the symbol R_{max} (see, Fig. 6.18). Because of the sensitivity of this parameter to scratches, it is usual to take the mean (R_{tm}) of the individual R_{max} of a number (normally five) of consecutive sampling lengths as shown in Fig. 6.18 (Dagnall 1980).

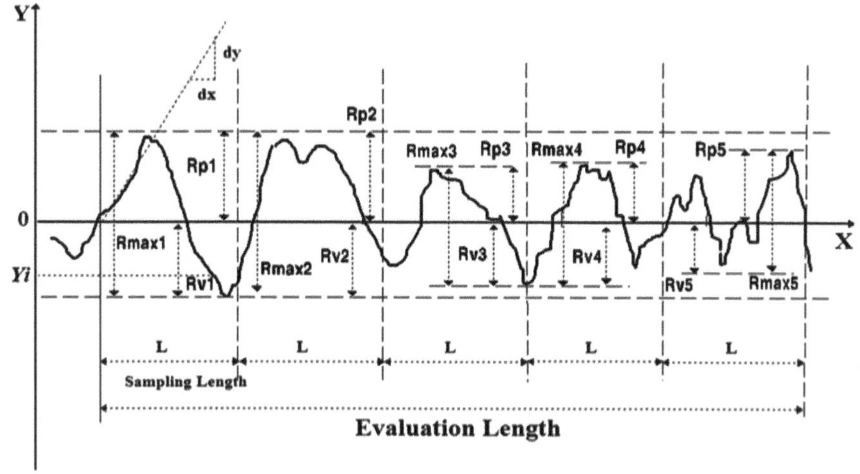

Fig. 6.18 A schematic illustration on the surface profile (Kim and Smith 2000)

The R_{tm} is the average of the maximum peak-to-valley height (R_{max}) of five consecutive sampling lengths (SL). But the R_t is the maximum peak-to-valley height of the profile within the assessment length (AL). Both parameters of the R_{tm} and R_{pm} indicate the maximum height of roughness profile, but there is an also difference between them (see, Fig. 6.18).

$$R_{tm} = \frac{R_{max1} + R_{max2} + \cdots + R_{max5}}{5} \tag{6.13}$$

$$R_{pm} = \frac{R_{p1} + R_{p2} + \cdots + R_{p5}}{5} \tag{6.14}$$

$$R_{vm} = \frac{R_{v1} + R_{v2} + \cdots + R_{v5}}{5} \tag{6.15}$$

According to the above definition, R_{tm} is the average of peak to valley height in each cut-off length and is equal to the sum of R_{pm} and R_{vm}. That is,

$$R_{tm} = R_{pm} + R_{vm} \tag{6.16}$$

In addition to the above-mentioned height-related parameters, the asperity slope is often analysed in connection with friction and wear properties (Sherrington and Smith 1987; Woo and Thomas 1979).

Before initial sliding of the shoe heel against the flooring counterface has occurred, the total friction force required for sliding can be considered as the tangential force required to overcoming the adhesion at regions of intimate contact plus the tangential force required to lift the asperities over each other.

During this process, the angle of each asperity would directly affect the extent of tangential forces applied in the sliding direction Therefore, the asperity slope could be an important factor in frictional behaviour. The absolute average slope at a point, ***AASi***, is found as below (Nowicki 1985):

$$AAS_i = \frac{1}{n}\sum_{i=1}^{n}\left|\frac{dy}{dx}\right| \tag{6.17}$$

6.8.3.3 Shape Parameters

It is inevitable that in attempting to describe a surface roughness profile by a single number, some important information about the surface topography would be lost. As an example, both R_a and R_q roughness parameters provide no information on the shapes or spacing of the surface irregularities and convey no indication of the probability of finding surface heights within certain limits.

For a full description of the surface topography, therefore, further information is needed to the probability distribution of surface heights and the spatial distribution of peaks and valleys across the surface. The need for a method of describing the distribution of surface heights is met by defining an amplitude density function, $P(y)$, which is proportional to the probability of finding a point on the surface at a height y above the mean line. The quantity $p(y)\Delta y$ is then the fraction of the surface profile which lies at heights between y and $y + \Delta y$ above the mean line.

An asymmetrical roughness profile leads to an amplitude density curve which is symmetrical about the position of the mean line. Whilst asymmetry of the surface profile leads to skewing of the amplitude density function, it contains information on the shapes of surface irregularities as well as their vertical extent.

Parameters which describe the shape of a surface add an extra dimension to the analysis of surface characteristics. The shape of the amplitude density curve can be described by skewness (R_{sk}), which measures its asymmetry, and kurtosis (R_k), which measures sharpness of the peaks of the distribution curve. Hence, the distribution of heights can be described by numerical values. R_{sk} distinguishes between asymmetrical profiles which could exhibit the same R_a value.

These shape parameters may provide a visual interpretation of surface profiles more easily. Skewness and Kurtosis have been shown to be of value in studies of running-in wear (King et al. 1977) and analyses of surfaces which attempt to relate profile characters and functional requirements of surfaces (Stout and Davis 1984). The skewness and the kurtosis are defined by

$$R_{sk} = \frac{1}{N \times R_q^3} \times \sum_{i=1}^{N} Y_i^3 \tag{6.18}$$

$$R_k = \frac{1}{N \times R_q^4} \times \sum_{i=1}^{N} Y_i^4 \quad (6.19)$$

A useful reference for R_{sk} and R_k is the case where their values for a surface have a normal distribution of surface heights. In this unique condition, R_{sk} is zero and R_k are 3. If $R_k < 3$, the surface profile has relatively few narrow spikes. As R_k increases above 3, the shape of its peaks and valleys become less rounded.

Many surface generation processes such as lapping and plateau honing produce asymmetric surfaces which generally tend to be negative skewed (Stout and Davis 1984). The direction of skew indicates whether the bulk of the material is above (negative skew) or below (positive skew) the mean line Fig. 6.19 shows the difference between positive and negative skewness. Negative skew is considered characteristic of a good bearing surface, indicating the presence of comparatively few spikes, which should wear away quickly. Figure 6.20 further illustrates skewness and kurtosis in the variable distributions.

On the other hand, although a surface with positive skew may acquire an adequate bearing face, it is likely to retain lubricant poorly. Positive skew is sometimes specified for electrical contacts because even a light contact load creates enough pressure to deform a few protruding peaks to the point of cracking an inelastic and insulating oxide film (Schaffer 1988).

Kurtosis is a measure of the peakedness or spikiness of a surface profile and may be relevant in the assessment of surfaces which are subject to sliding friction. A feature of the skewness and kurtosis parameters is that they are interrelated numerically and therefore it is useful to associate a high numerical value of the skewness, either positive or negative, with a high numerical value of the kurtosis (see, Fig. 6.20).

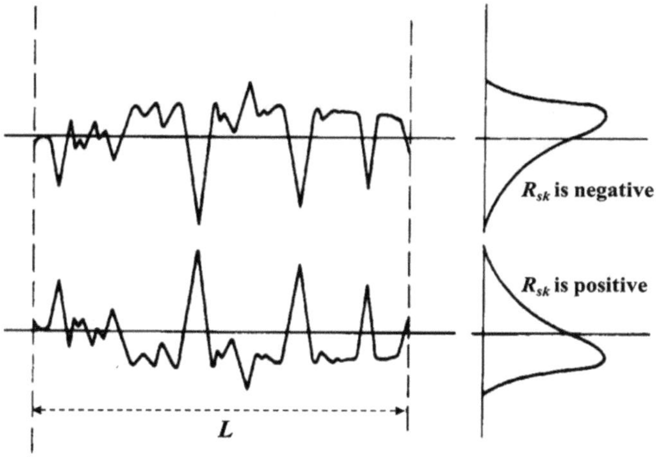

Fig. 6.19 Schematic illustration of the difference between positive and negative skewness

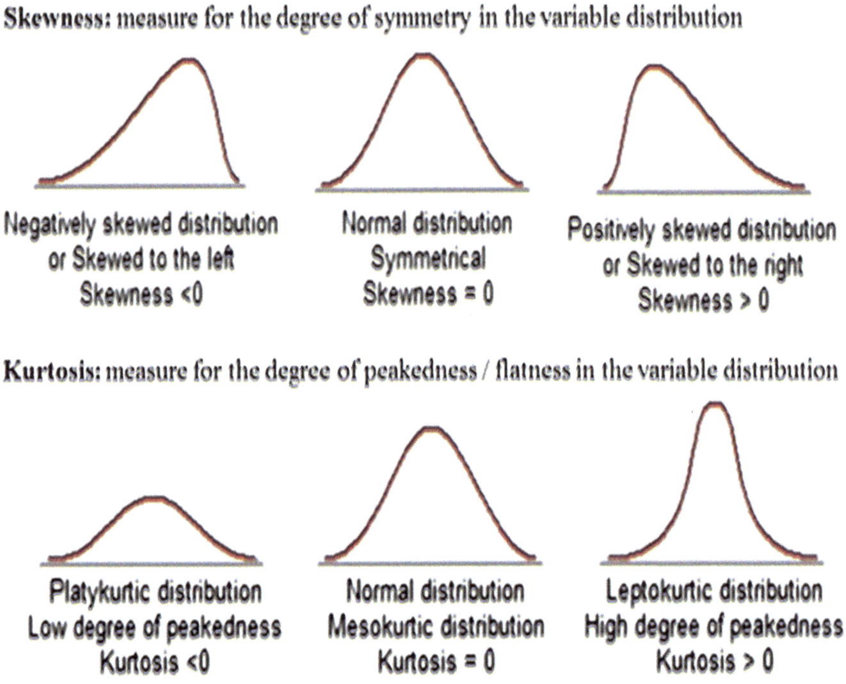

Fig. 6.20 Schematic illustrations on the Skewness and Kurtosis (Shawn's Statistics Tutoring, 2016)

6.8.4 Spatial Distribution of Surface Texture

Several roughness parameters have been formulated to describe spatial characteristics of surfaces. They have not been included as standard instrumental readings until the last three decades and this may account for their use being less widespread than height parameters such as $\boldsymbol{R_a}$ and $\boldsymbol{R_q}$ (Sherrington and Smith 1987). In the absence of standardisation, following definitions for waviness parameters are based on the studies of Nowicki (1985) and Schaffer (1988).

6.8.4.1 Average Wavelength

An average wavelength, $\boldsymbol{\lambda a}$, is defined by

$$\lambda_a = 2\pi \frac{\boldsymbol{R_a}}{\Delta_a} \tag{6.20}$$

where Δ_a is the average slope of a profile.

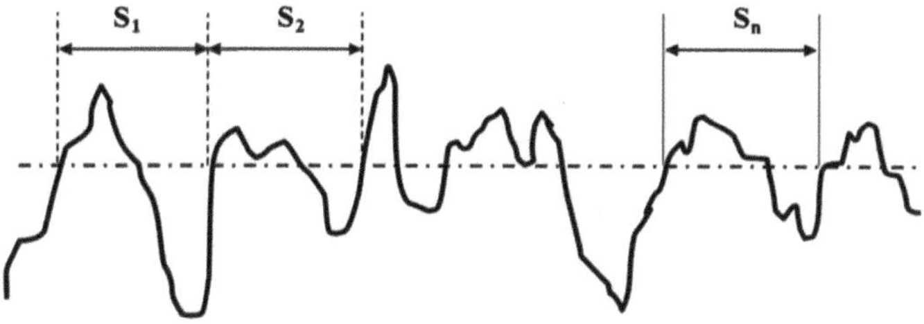

Fig. 6.21 Graphical interpretation of the average wavelength. It shows mean spacing of the profile irregularities

Figure 6.21 shows graphical shapes of the average wavelength. It measures the spacing between local peaks and valleys. This parameter is useful for assessing the surface quality. It is able to measure the 'openness' of a surface and is found to correlate well with its visual appearance (Sherrington and Smith 1987). This parameter is a sensitive detector for the feed marks of a machine tool (Dagnall 1980).

6.8.4.2 RMS Wavelength

A root mean square (RMS) wavelength, λq, is defined by

$$\lambda q = 2\pi \frac{R_q}{\Delta_q} \tag{6.21}$$

where Δ_q is the root mean square slope of a profile. λq has similar attributes to λa discussed above.

6.8.4.3 High Spot Count (HSC)

Peaks in a surface profile are frequently important from a functional viewpoint. It has been reported that the spacing of roughness peaks exerts an important influence (Dagnall 1980). Normally, high spot count (HSC) is assessed by determining the number of peaks within a sampling length. However, several different definitions of a 'peak' are encountered. Most commonly, HSC is determined by counting the number of excursions per unit length above a profile mean line (Thomas 1981).

6.8.5 Hybrid Parameters

A small number of roughness parameters are in use which does not measure solely amplitude or spatial characteristics of roughness profiles but serve to assess other familiar topographic characteristics.

6.8.5.1 Mean Slope (Δa and Δq)

A profile slope is an angle that it makes a line parallel to the centre line. The mean of the slopes at all points in the profile within the sampling length is known as the average slope (symbol Δ_a for the arithmetic mean or Δ_q for the RMS value) (Nowicki 1985; Garmo et al. 1997).

This parameter was suggested as long as 1962 by Myers (1962) and one example of its use is to measure the developed or actual profile length. That is, the length occupied if all peaks and valleys are stretched out in a straight line, the distance one would have to walk up and down dale to traverse all the peaks and valleys (see, Fig. 6.22). The steeper the average slope, the longer the actual length (***L'***) of the surface compared with its nominal length (***L***).

The sliding interface between the shoe heel and floor surface could be regarded as a tangential force required to overcome the adhesion at regions of intimate contact plus the tangential force required to lift the asperities over each other.

During this process, the angle of each asperity would significantly affect the extent of tangential forces applied in the sliding direction Therefore, the average slope of asperities needs to be involved in investigating its role of the friction results between the shoe heel/sole and floor surface.

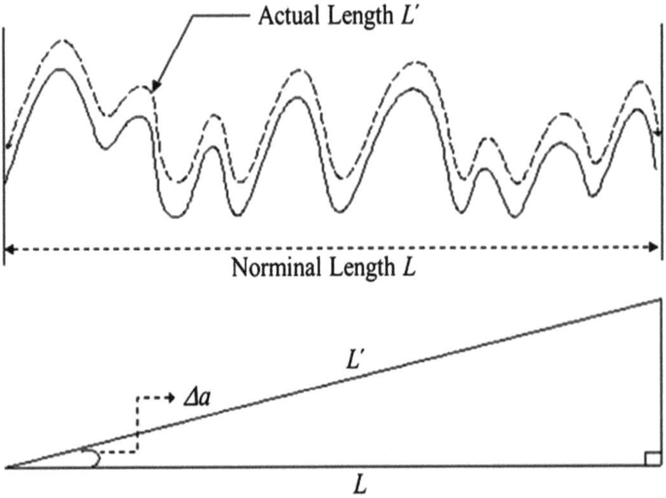

Fig. 6.22 A schematic diagram for estimation of the true length L′ of a surface profile

The average slope Δ_a of a profile *f(y)* is given by (Spragg and Whitehouse 1970)

$$\Delta_a = \frac{1}{L}\int_0^L \left|\frac{df(y)}{dy}\right| dy \tag{6.22}$$

6.9 Relationships Amongst Surface Roughness Parameters

When the distribution of ordinate heights in a surface profile is Gaussian, a relationship between the R_a and R_q parameters can be obtained and an estimate of one parameter calculated from a knowledge of the other (Sherrington and Smith 1987). Such relationships can be derived for many theoretical height distributions.

King and Spedding (1982) have investigated the relationships amongst a wide range of roughness parameters and shown that knowledge of R_a, R_q, R_{sk}, and R_k parameters defines the shape of a certain class of height distribution curves of the ordinates (that is, Pearson & Johnson distributions) sufficiently well to allow a number of well-known parameters to be estimated reliably from them.

The skewness and kurtosis parameters have also been used to estimate parameters of the bearing area curve (King and Spedding 1982). Definitions for the non-graphical presentation of bearing area curve data given in the study of Abbott and Firestone (1933) were adopted. Peak roughness was defined as the range of heights which covered 2–25% of the bearing area, median roughness as that range which enclosed 25–75%, and valley roughness as the range containing 75–98% of the bearing area.

On the other hand, extreme roughness parameters such as R_{max}, R_t, and R_{tm} cannot be estimated from the knowledge of height distribution alone as they depend on spatial considerations for a given length of the roughness profile. The interdependence of surface roughness parameters is a complex situation and beyond this book again.

The above facts indicate that it seems to be possible to use a statistical approach to estimate unknown parameters of a surface roughness profile. However, estimating roughness parameters which describe the surface as a whole from the details of roughness profiles is a more difficult task. This is because different statistical relationships hold between profile parameters and surface parameters, or between parameters evaluated from profiles recorded from different parts of a specimen (Sherrington and Smith 1987).

In addition to the above reality, during the repetitive friction process, the initial surface profiles of both shoe and floor are continuously changing through various wear behaviours and interactive wear mechanisms. As a result, estimating roughness parameters from others seem to distort the overall quality of data themselves. As also pointed out by Sherrington and Smith (1987), in most practical circumstances such factors as inhomogeneity in the quality of the work material, tool wear and damage, and presence of cutting debris exert a significant effect. However, one of the most serious problems associated with the surface measurement and analysis is to obtain consistent data.

In order to acquire coherent data and to maintain constant measuring results, it is necessary to standardise the measurement conditions such as measuring the bandwidth of the profile, sampling interval, filtering, and measuring times. In this book, a newly developed laser scanning confocal microscopy was adopted to measure the surface roughness profiles of shoe and floor specimens.

6.10 Surface Analysis for the Shoe-Floor Friction System

To explore topographic changes, surface finishes of the floor and shoe should be thoroughly investigated with suitable surface roughness parameters, and their surface alterations need to be analysed. Three-dimensional analyses for the worn surfaces of shoes and floors may provide more extensive information for the qualitative and quantitative characterizations of both bodies than the conventional ones. The contribution of surface roughness to slip resistance performance is linked to the fundamental knowledge of friction and wear behaviours between the shoe and floor surface (Kim and Nagata 2008a, b; Kim et al. 2013; Kim 2015a, c, 2016a, b).

Detailed studies on both surfaces of the shoe and floor would provide vital information to understand the complex features of friction and wear behaviours and mechanisms and their interactive effects on slip resistance performance. However, these factors have rarely been considered so far and simple friction measurements have commonly practised.

As pointed out in the previous chapters, this practice is mainly caused by underestimating the complex nature of tribo-physical characteristics of friction and wear activities of shoes and floors and their joint effects on slip resistance performance. Special caution is required for the determination of slip resistance properties under a range of walking environments. Facilitated routine friction measurements in laboratory conditions could also oversimplify the essential characteristics of slip resistance properties between the floor and shoe surfaces.

It is not yet known which method of measuring COFs best represents the pedestrian slip resistance properties. To date, no comparative studies have shown which method of measuring surface roughness or which roughness parameter(s) produces the strongest correlation with the COF for a given shoe-floor combination during the human ambulation.

It also should be emphasised that further research to comparing the results of various methods for measuring surface roughness is necessary, but cannot be conclusive until the method of measuring COF is agreed internationally. Although there has been considerable progress to understand the parameters of surface roughness that correlates with the COFs, it is probably true to say that none of the COF measurements under polluted environments reported to date can be regarded as final objective values or quantities for a specified floor-shoe-contaminant combination.

The recent literature review showed that most of the studies on surface roughness measurements were related to slipperiness through friction measurements. As friction measurements related to slipperiness remain controversial, however, the problems with friction measurement can overshadow the contribution of research on surface roughness to slipperiness measurements. Therefore, it is essential for researchers working on surface roughness and surface analysis to find some other evaluation methods to assess the slipperiness for complementing the current activities. Other methods used to measure slipperiness may include subjective and biomechanical methods such as the measurement of slip distance and sliding velocity.

However, as explored in this chapter, surface roughness seems to be a critical factor that can significantly affect the slip resistance performance between the shoe and floor surface although it is not the only factor. Thus, future studies require to develop a solid theory concept and model and establish national and international database that can allow safety researchers, practitioners and industry shoe and floor manufacturers to access needed information on the surface roughness of shoes and floors as a possible indicator for the slipperiness assessment.

6.11 Development of a Contact Model Between the Shoe and Floor Surface

6.11.1 Introduction

As the main material property of the footwear, polymers (or elastomers) have been known that they consist of long-chain organic molecules linked at various points along their length by strong chemical bonds, the elastic modulus increasing with the number of cross-linking bonds (Cohen and Tabor 1966). It has been observed that when a polymer adheres to a hard counterface, the pull-off work is proportional to the thermodynamic work of adhesion but very much greater. The high values of pull-off work are also associated with deformation losses, but in this case, the losses are due to the flow of the polymer rather than hysteresis losses (Cohen and Tabor 1966). The pull-off force is also critically dependent on the surface roughness of the counterface and decreases rapidly as the roughness increases.

The friction mechanism of polymers involves both adhesion and deformation damages, but there is little or no junction growth. Also, the deformation losses during sliding friction events are much more dependent on the surface roughness of the counterface. Polymers generally have a low melting or softening point and poor thermal conductivity, so when there is significant frictional heating, the surface layer usually softens or melts (Kim et al. 2001). The friction then largely arises from viscous losses in the surface layers (Moore and Geyer 1972). At low sliding speeds, where frictional heating is negligible, the frictional characteristics depend on the roughness of the counterface (Roth et al. 1942; Fuller and Tabor 1975; Lancaster 1988). Therefore, it becomes clear that the role of surface roughness of both shoes and floors is s critical factor to observe the shoe-floor slip resistance properties.

6.11.2 Main Hypotheses for Contact Model Development

The concept on changes in surface topographies of two interacting bodies as a result of sliding friction action can be applied to the sliding interface between the shoe and floor. The following is a summary of main assumptions for the development of a theory model on the contact-sliding mechanism between the shoe heel and floor surface:

(1) Firstly, it is considered that a contact mechanism between the shoe and floor surface is elastoplastic deformations (see, Fig. 6.23). The deformations are assumed to be mainly concentrated on the shoe surface, whose elastoplastic modulus is significantly less than that of the floor surface.
(2) When a shoe heel is loaded against a floor surface, it could be assumed that two surfaces initially touch only at tiny discrete areas where their highest asperities are in contact, as illustrated in Fig. 6.23 (stage 1). A local pressure at the contact regions would be high enough to cause plastic deformation of the heel asperities even at the lightest load due to an elastic modulus of the shoe heel which is considerably less than that of the floor surface. Hence, its contact mechanism would be an interlocking status as shown in Fig. 6.23 (stage 2). If the shoe heel repeatedly slides on the floor surface, then the heel surface would be ruptured, deformed and ploughed by the wedge-shaped hard asperities of the floor surface (see, stage 3 in Fig. 6.23).
(3) It also can be assumed that tribological behaviours of the shoe-floor pair are greatly influenced by surface topographies of the floor counterface. This concept can be idealised as a contact-sliding model between the shoe heel and floor surface. That is, the shoe heel slides over an array of wedge-shaped hard asperities of the floor surface. As the shoe heel touches the floor surface, the high asperities of the floor surface penetrate into the shoe heel areas and make real areas of contact.
(4) If the shoe slides on the floor surface, the shoe surface will be ruptured and deformed by wedge-shaped asperities of the floor surface. From this conceptual model, it can be considered that the density of peak height (denseness of peak asperity within the assessment length) of the floor surface seems to be a major factor in affecting wear developments of the shoe surface. In this process, an asperity attack angle (θ) of the wedges may play an important role in the configuration of the shoe surface deformation. Therefore, the average asperity angle of the floor surface needs to be investigated.

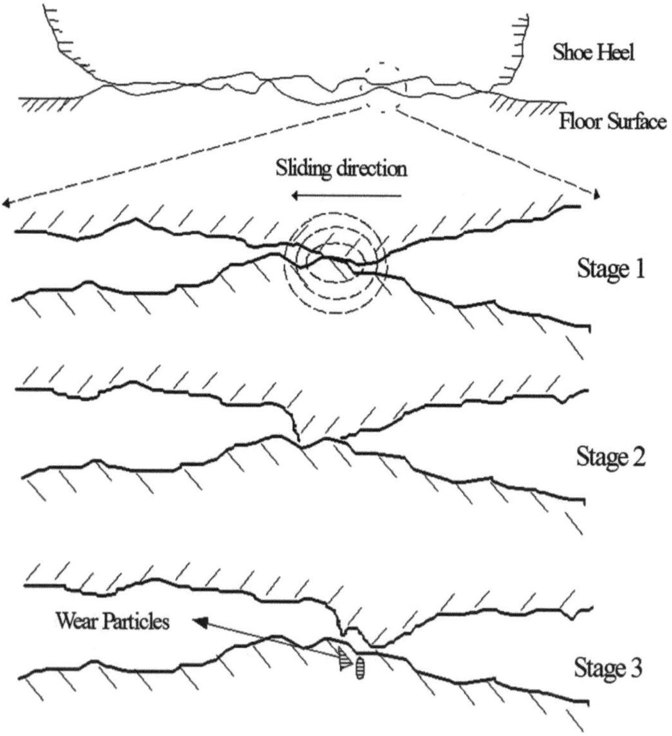

Fig. 6.23 Schematic illustration of a unit event in the geometrical interaction between a shoe heel and floor surface during dry friction processes. *Stage 1* elastic and plastic deformations and ploughing. *Stage 2* adhesion bonding between the shoe heel and floor surface. *Stage 3* shearing, plastic deformation, and wear

(5) During the recurring sliding friction events, topographies of the floor surface also can be affected by several causes. Amongst various possible reasons, deposition of abraded polymer products from the heel surface into the crevices of asperities on the floor surface seems to be one of the most important considerations. This means that valley areas of the floor asperities can be one of the vital parameters. Hence, this factor requires thorough investigations and continuous observations. These aspects clearly indicate that a contact-sliding friction mechanism between the shoe and floor surface significantly depends on the surface topography of the floor counterface.

Based on the above assumptions, Fig. 6.24 suggests a contact-sliding model of the sliding interface between the shoe heel and floor surface during the repetitive sliding friction processes.

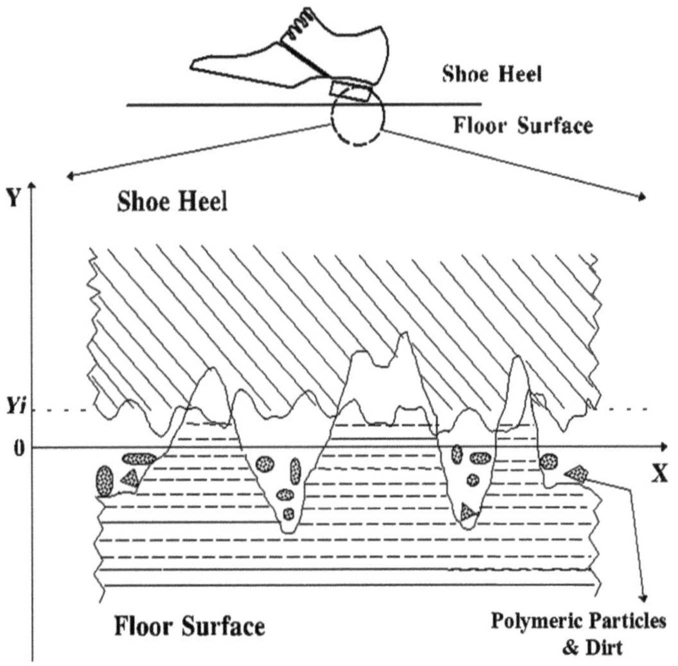

Fig. 6.24 A schematic diagram for a contact-sliding model proposal between a shoe and floor surface

6.11.3 Model Development

When a shoe heel is loaded against a floor counterface, initial contact seems to be made of the peak areas of a relatively small number of asperities identified as ***A*** to ***D*** in Fig. 6.25. Because the elastic modulus of the floor surface is so much greater than that of the shoe heel, peak asperities of the floor counterface seem to penetrate into the heel surface.

In this condition, it can be assumed that both initial surfaces of the shoe and floor possess Gaussian distributions of asperities, respectively. These height distributions are also shown diagrammatically in Fig. 6.25.

In the above model, mutual influences between the two surfaces can be considered as a manner of normal and frictional loading. When gross sliding occurs, the heel surface seems to undergo both normal and horizontal displacements. This can be explained by assuming that when the heel surface is displaced from the rest it must climb up and pass the forward of asperity slopes.

From this condition, following two key components are considered separately (Kim and Nagata 2008b).

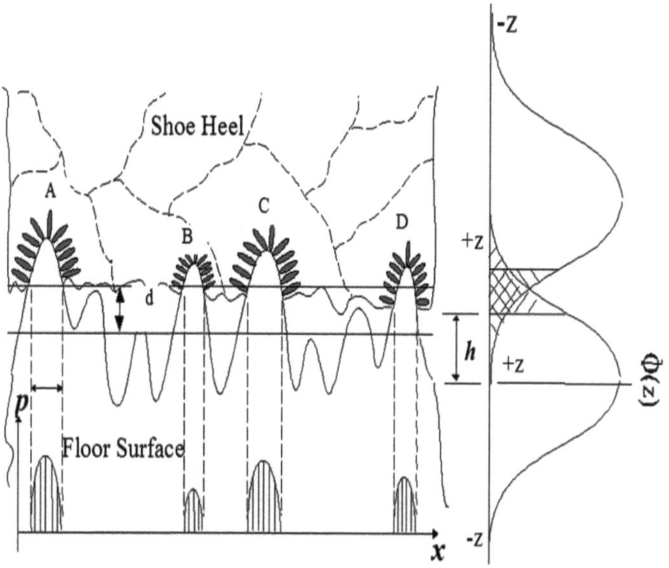

Fig. 6.25 A schematic diagram for a contact model between a shoe heel and a floor surface with Gaussian height distributions (Kim and Nagata 2008b)

6.11.3.1 Consideration of Normal Loading

The total normal load assumes to be found in the same way as by Greenwood and Williamson (1966). When two rough Gaussian surfaces of the shoe and floor are brought into normal dry contacts, there may be a gap, ***d***, between their reference planes, and asperities with heights originally greater than ***d***, seem to be in contact each other. In this topographical situation, the load carried by an individual asperity seems to be a function of its compression:

$$W_I = f_p(z_i - d) \tag{6.23}$$

The number of asperities with heights between z and $z + dz$ is $AnDa\varphi(z)dz$, and the total load for a separation d thus becomes:

$$W = AnDa \int_d^{\infty} f_p(z_i - d)\varphi(z)dz \tag{6.24}$$

where ***An*** is the sum of each of the discrete areas ***Ai*** and ***Da*** is the density of contour.

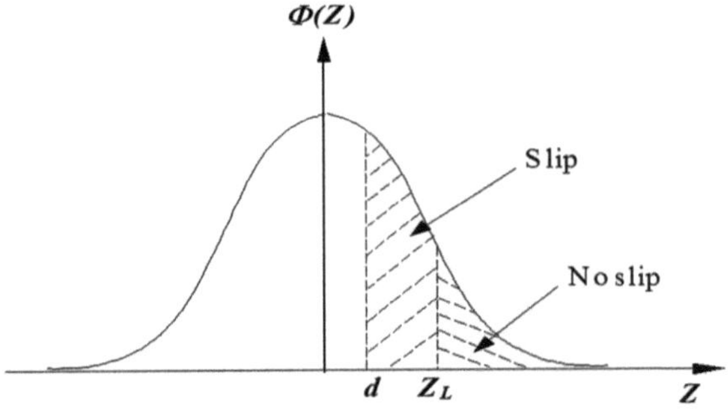

Fig. 6.26 Schematic representation of asperity height distribution with the region of slip and stick (Kim and Nagata 2008b)

6.11.3.2 Consideration of Frictional (or Tangential) Loading

The normally loaded contact is subsequently subjected to a tangential load, applied by moving the shoe surface horizontally a distance δ. An individual asperity can now behave in two different ways depending on its original height (see, Fig. 6.26). An asperity lower than a limiting height z_L seems to slip and therefore contributes to the total frictional (or tangential) force with $F_i = \mu W_i$. An asperity higher than z_L may exhibit partial slip, but has some part of its contact area that is sticking, and is carrying a tangential force that can be a function of the global tangential displacement and the normal compression:

$$F_i = f(\delta, z_i - d) \tag{6.25}$$

which is valid for loads less than the limiting friction, that is, $F_i \leq \mu W_i$.

When gross sliding occurs during unlubricated contacts, the upper body (shoe heel surface) may undergo both normal and horizontal displacements. This can be explained by assuming that when the shoe surface is displaced from the rest it must climb up and pass the forward of asperity slopes. That is, the total frictional force is given by

$$F = F_{adhesion} + F_{deformation} \tag{6.26}$$

The above theoretical concepts can be used as a basis for a model formulation with the consideration of friction and wear behaviours of the shoe and floor and their interactive effects on slip resistance performance.

6.12 Conclusions

To walk safely without falls, it is important to maintain an adequate level of slip resistance or traction properties between the shoe sole/heel and floor surface. It is universally recognised that slippery surfaces are dangerous for safe ambulation. The roughened floor surface has a beneficial effect in raising COFs significantly between the shoe heel/sole and floor surface under polluted environmental conditions.

As clearly demonstrated in this chapter, there are convincing evidences that surface roughness of the shoe and floor extensively affect slip resistance properties and is significantly correlated with slipperiness assessment through friction measurements. As the controversies around the friction measurements still remain, improvements in the methodologies for assessing slip resistance behaviours and expansions into other methods and concepts on tribo-physical characteristics such as surface alterations are urgently needed to explore.

Also, the surface roughness of both shoes and floors should be measured and combined to arrive at the final assessment of slipperiness. Therefore, surface features of the shoe heels and floor surfaces should be monitored routinely to select and sustain the best shoe-floor combinations for specific walking environments. This may provide a more reliable result for monitoring pedestrian fall safety than considering the friction measurement alone.

In conclusion, rather than simply measuring and specifying a minimum value of COF, considerations should be focused on establishing principal approaches for the investigation of complicated slip resistance properties. Future research should be based on a meticulous understanding of the nature of friction and wear behaviours and related tribo-physical elements, and surface analysis models and techniques for describing both surfaces of the shoe and floor and their interactions during repeated sliding friction processes. Therefore, tribological approaches would be a worthwhile way to overcome limitations on the existing researches and practices for the pedestrian fall safety measurements and assessments.

6.13 Chapter Summary

The measurement and interpretation of slip resistance properties should be based on full understanding of the relevant tribo-physical characteristics and involved mechanisms as an essential prerequisite because the friction measurement amongst the shoe, floor, and environment are not a simple matter. In this sense, it would be a constructive attempt to study the topographic features of surfaces, their contact-sliding mechanisms, and related tribo-physical behaviours.

In order to recognise tribological processes involved at the sliding interface between the shoe and the floor surface, we must understand how the two surfaces interact when they are loaded together. Surface analyses and relevant background information were comprehensively reviewed with the measuring instruments to quantify topographic characteristics of the shoe and floor surfaces in this chapter.

Several geometrical models were comprehensively reviewed and related theoretical issues were also carefully discussed. The concept on changes in surface topographies of two interacting bodies as a result of sliding friction action was applied to the sliding interface between the shoe and floor. Based on the main assumptions with thorough knowledge foundations for the contact-sliding mechanisms between two surfaces, a theory model on the contact-sliding mechanism between the shoe heel and floor surface was suggested. A perfect contact-sliding model to predict slip resistance performance between the shoe and floor surface has not developed yet, but tribological approaches seem to be a worthwhile attempt to overcome limitations on the existing researches and practices for the measurements of pedestrian fall safety.

References

Abbott, E. J., & Firestone, F. A. (1933). Specifying surface quality—A method based on accurate measurement and comparison. *Journal of Mechanical Engineering*, (55), 569–572.

Abouelatta, O. B., El-Bediwi, A., Sakrana, A., Jiang, X. Q., & Blunt, L. (2006). Surface integrity of provisional resin materials. *Measurement Science & Technology, 17*(3), 584–591.

Ahmed, R., Hadfield, M. (1999). Failure modes of plasma sprayed WC-15%Co coated rolling elements. *Wear, 230*(1), 39–55.

Anamalay, R. V., Kirk, T. B., & Panzera, D. (1995). Numerical descriptors for the analysis of wear surfaces using laser scanning confocal microscopy. *Wear*, 181–183, Part 2, 771–776.

Arnell, R. D., Davies, P. B., Halling, J., & Whomes, T. L. (1991). Surface properties and surface contact. In *Tribology—Principles and design applications* (pp. 1–26). New York: Macmillan Education, Ltd.

Bhushan, B. (2000). Surface roughness analysis and measurement techniques. In *Modern tribology handbook*, Two Volume Set (Chapter 2) Boca Raton: CRC Press.

Bhushan, B., Wyant, J. C., & Meiling, J. (1988). A new three-dimensional digital optical profiler. *Wear, 122*(3), 301–312.

Bowden, F. P., & Tabor, D. (1954). *The friction and lubrication of solids* (Chapters 1 and 2). Oxford: Oxford Clarendon Press.

Chang, W. R., Kim, I. J., Manning, D. P., & Bunterngchit, Y. (2003). The role of surface roughness in the measurement of slipperiness. In *Measuring slipperiness—Human locomotion and surface factors* (pp. 101–117). London: Taylor & Francis.

Cohen, S. C., & Tabor, D. (1966). The friction and lubrication of polymers. *Proceedings of Royal Society, B66,* 721.

Dagnall, H. (1980). *Exploring surface texture*. Leicester, England: Rank Taylor Hobson.

Dieter, G. E. (1986). *Mechanical metallurgy* (3rd ed.). New York, USA: McGraw-Hill Education.

Dong, W. P., Sullivan, P. J., & Stout, K. J. (1994). Comprehensive study of parameters for characterising three-dimensional surface topography: III: Parameters for characterising amplitude and some functional properties. *Wear, 178*(1–2), 29–43.

Dowson, D. (1998). *History of tribology* (2nd Ed.). Professional Engineering Publishing (Distributed by ASTM).

Fellers, T. J., & Davidson, M. W. (2007). *Introduction to confocal microscopy*. Olympus Fluoview Resource Center. National High Magnetic Field Laboratory.

Flinn, R. A., & Trojan, P. K. (1975). *Engineering materials and their applications* (p. 61). Boston: Houghton Mifflin Company.

Fuller, K. N. G., & Tabor, D. (1975). The effect of surface roughness on the adhesion of elastic solids. *Proceedings of Royal Society, A345,* 327–342.

Garmo, E. P., Black, J. T., & Kosher, R. A. (1997). *Materials & processes in manufacturing* (8th ed.). New York: Macmillan, Prentice-Hall.

Greenwood, J. A., & Williamson, J. B. P. (1966). Contact of nominally flat rough surfaces. *Proceedings of Royal Society, London, A295*(1442), 300–319.

Grönqvist, R. (1995). Mechanisms of friction and assessment of slip resistance of new and used footwear soles on contaminated floors. *Ergonomics, 38*(2), 224–241.

Grönqvist, R., Roine, J., Järvinen, E., Korhonen, E., & Rahikainen, A. (1990). Slip resistance versus surface roughness of deck and other surfaces in ships. *Journal of Occupational Accidents, 13*(4), 291–302.

Halling, J. (1976). *Introduction to tribology*. London: Wykeham.

Harris, G. W., & Shaw, S. R. (1988). Slip resistance of floors: Users' opinions, Tortus instrument readings and roughness measurement. *Journal of Occupational Accidents, 9*(4), 287–298.

Hillmann, W., Kranz, O., & Eckolt, K. (1984). Reliability of roughness measurements using contact stylus instruments with particular reference to results of recent research at the physikalisch-technique bundesanstalt. *Wear, 97*(1), 27–43.

Holmberg, K., & Matthews, A. (2000). Tribological properties of metallic and ceramic coatings. In Bhushan, B. (Ed.), *Modern tribology handbook*, Two-Volume Set (Chapter 23). Boca Raton: CRC Press. Available from: http://home.ufam.edu.br/berti/nanomateriais/8403_PDF_CH23.pdf

Hutchings, I. M. (1992). *Tribology—Friction and wear of engineering materials* (p. 23). Metallurgy and Materials Science Series. London: Edward Arnold.

Inoue, T. (1990). The relationship of sea surface temperature and water vapor amount to convection over the western tropical pacific revealed from split window measurements. *Journal of the Meteorological Society of Japan, 60*(5), 589–606.

ISO 3274. (1996). Geometrical Product Specifications (GPS)—Surface texture: Profile method. Nominal characteristics of contact (stylus) instruments.

Johnson, K. L. (1985). *Contact mechanics*. Oxford, Clarendon: Cambridge University Press.

Kang, K., Pelow, C., & Witham, L. (2008). Analysis of changes in 3D surface texture anisotropy ratio parameter and friction during sheet rolling campaigns. 10th International Conference on Metrology and Properties of Engineering Surfaces. *Wear, 264*(5–6), 434–438.

Kim, I. J. (1996a). Tribological concepts for the investigation of the pedestrian slipping and falling accidents—Part I. International Occupational Injury Symposium (p. 101), February, Sydney, Australia.

Kim, I. J. (1996b). Tribological approach for the analysis of pedestrian slip hazard—II. Proceedings of the '96 Spring Conference of Korean Institute of Industrial Engineers (pp. 279–285), April, Seoul, Korea.

Kim, I. J. (1996c). Microscopic investigation to analyze the slip resistance of shoes. Proceedings of the 4th Pan Pacific Conference on Occupational Ergonomics (pp. 68–73), November, Taiwan, ROC.

Kim, I. J. (1996d). Microscopic observation of shoe heels for pedestrian slip hazard investigation. Proceedings of the 1st Annual International Conference on Industrial Application and Practice (pp. 243–250), December, Texas, U.S.A.

Kim, I. J. (2004a). Development of a new analyzing model for quantifying pedestrian slip resistance characteristics: Part I—Basic concepts and theories. *Industrial Journal of Industrial Ergonomics, 33*(5), 395–401.

Kim, I. J. (2004b). Development of a new analyzing model for quantifying pedestrian slip resistance characteristics: Part II—Experiments and validations. *Industrial Journal of Industrial Ergonomics, 33*(5), 403–414.

Kim, I. J. (2005a). Combined effects of friction and wear behaviors on fall safety measures. In *Contemporary ergonomics* (pp. 498–502). London: Taylor & Francis.

Kim, I. J. (2005b). A new understanding on the shoe wear mechanism and its significance on slip resistance property. In *Contemporary ergonomics* (pp. 503–508). London: Taylor & Francis.

Kim, I. J. (2006a). The current hiatus in fall safety measures. In W. Karwowski (Ed.), *International encyclopedia of ergonomics and human factors-2005* (pp. 2572–2576). Philadelphia, USA: Taylor & Francis Group, LLC.

Kim, I. J. (2006b). A new paradigm for characterizing slip resistance properties. In W. Karwowski (Ed.), *International encyclopedia of ergonomics and human factors-2005* (pp. 2735–2740). Philadelphia, USA: Taylor & Francis Group, LLC.

Kim, I. J. (2015a). Practical design search for optimal floor surface finishes to prevent fall incidents. In B. Evans (Ed.), *Accidental falls: Risk factors, prevention strategies and long-term outcomes* (Chapter 5, pp. 80–103). Hauppauge, NY, USA: Nova Science Publishers, Inc.

Kim, I. J. (2015b). Slip-resistance measurements for assessing pedestrian falls: facts and fallacies. In B. Evans (Ed.), *Accidental falls: Risk factors, prevention strategies and long-term outcomes* (Chapter 6, pp. 105–125). Hauppauge, NY, USA: Nova Science Publishers, Inc.

Kim, I. J. (2015c). Wear observation of shoe surfaces: Application for slip and fall safety assessments. *Tribology Transactions, 58*(3), 407–417.

Kim, I. J. (2015d). Research challenges on slip-resistance measurements for assessing pedestrian fall incidents. *Journal of Ergonomics, 5*(3). doi:10.4172/2165-7556.1000e142

Kim, I. J. (2016a). A study on wear development of floor surfaces: Impact on pedestrian walkway slip-resistance performance. *Tribology International, 95,* 316–323.

Kim, I. J. (2016b). Identifying shoe wear mechanisms and associated tribological characteristics: The importance for slip resistance evaluation. *Wear, 360–361,* 77–86.

Kim, I. J., Hsiao, H., & Simeonov, P. (2013). Functional levels of floor surface roughness for the prevention of slips and falls: Clean-and-dry and soapsuds-covered wet surfaces. *Applied Ergonomics, 44*(1), 58–64.

Kim, I. J., & Nagata, H. (2008a). Nature of the shoe wear: Its uniqueness, complexity and effects on slip resistance properties. In *Contemporary ergonomics* (Vol. 15, pp. 728–734). London: Taylor & Francis.

Kim, I. J., & Nagata, H. (2008b). Research on slip resistance measurements—A new challenge. *Industry Health, 46*(1), 66–76.

Kim, I. J., & Smith, R. (1998a). A study of the comparative geometry mating between the surfaces of the shoe and floor in pedestrian slip resistance measurements. The 5th Pan-Pacific Conference on Occupational Ergonomics (pp. 34–37), July, Kitakyushu, Japan.

Kim, I. J., & Smith, R. (1998b). Tribological characterization of the frictional force component in pedestrian slip resistance measurements. Third World Congress of Biomechanics (WCB '98), August, Hokkaido, Japan.

Kim, I. J., & Smith, R. (1999). The relationship between wear, surface topography characteristics and coefficient of friction as a means of assessing the slip hazards. 2nd Asia-Pacific Conference on Industrial Engineering and Management Systems (APIEMS'99) (pp. 155–161), October, Ashikaga, Japan.

Kim, I. J., & Smith, R. (2000). Observation of the floor surface topography changes in pedestrian slip resistance measurements. *Industrial Journal of Industrial Ergonomics, 26*(6), 581–601.

Kim, I. J., & Smith, R. (2001a). A critical analysis on the friction measuring concept for slip resistance evaluation. ASTM Symposium on the Metrology of Pedestrian Locomotion and Slip Resistance (pp. 1–14), June, ASTM Headquarters, West Conshodocken, Pennsylvania, USA.

Kim, I. J., & Smith, R. (2001b). A study for characterising topography changes of shoe surfaces in the early stage of slip resistance measurements—Bearing Area Curve. 6th Pan-Pacific Conference on Occupational Ergonomics (pp. 299–303), August, Beijing, P. R. China.

Kim, I. J., & Smith, R. (2001c). Three-dimensional analysis of floor surface wear during slip resistance measurements. 6th Pan-Pacific Conference on Occupational Ergonomics (pp. 304–308), August, Beijing, P. R. China.

Kim, I. J., & Smith, R. (2003). A critical analysis of the relationship between shoe-floor wear and pedestrian/walkway slip resistance. In M. I. Marpet & M. A. Sapienza (Eds.), *Metrology of Pedestrian Locomotion and Slip Resistance* (pp. 33–48), American Society of Testing and Materials, Special Technical Publication 1424, ASTM International, Philadelphia, USA.

Kim, I. J., Smith, R., & Nagata, H. (2001). Microscopic observations of the progressive wear on the shoe surfaces which affect the slip resistance characteristics. *Industrial Journal of Industrial Ergonomics, 28*(1), 17–29.

King, T. G., & Spedding, T. A. (1982). On the relationship between surface profile height parameters. *Wear, 83*(1), 91–108.

King, T. G., Watson, W., & Stout, K. J. (1977). Modelling the micro geometry of lubricated wear. In Proceedings Fourth Leeds-Symposium, Surface roughness effects in lubrication (pp. 333–343), Paper XII(iv). London: Mechanical Engineering Publications.

Kragelsky, I. V., & Alisin, V. V. (1981). Friction, wear, lubrication. In *Tribology Handbook* (Vol. 1). Moscow: Mir Publishers.

Kumagai, N., Sasajima, S., & Ito, H. (1978). "岩石の長年クリープ実験: 巨大試片約20年間・小試片約3年間の結果", (Long-term creep of rocks: Results with large specimens obtained in about 20 years and those with small specimens in about 3 years). Japan Society of Materials Science (Japan) (in Japanese). *Japan Energy Society*, *27*(293), 157–161.

Lancaster, J. K. (1988). Solid lubricants. In R. E. Booser (Ed.), *Theory and practice of tribology, Volume II: Theory and design* (Chapter 16, pp. 269–290). CRC Handbook of Lubrication. Boca Raton: CRC Press.

Leaver, R. H., Sayles, R. S., & Thomas, T. R. (1974). Mixed lubrication and surface topography of rolling contacts. *Proceedings of the Institution of Mechanical Engineers, 188*(1), 461–469.

Lim, C. Y. H., Lim, S. C., & Lee, K. S. (1999). Wear of TiC-coated carbide tools in dry turning. *Wear, 225–229*, Part 1, 354–367.

Manning, D. P., & Jones, C. (1994). The superior slip-resistance of footwear soling compound T66/103. *Safety Science, 18*(1), 45–60.

Manning, D. P., Jones, C., Rowland, F. J., & Roff, M. (1998). The surface roughness of a rubber soling material determines the coefficient of friction on water-lubricated surfaces. *Journal of Safety Research, 29*(4), 275–283.

McCool, J. I. (1984). Assessing the effect of stylus tip radius and flight on surface topography measurements. *Journal of Tribology, 106*(2), 202–209.

Moalic, H., Fitzpatrick, J. A., & Torrance, A. A. (1987). The correlation of the characteristics of rough surfaces with their friction coefficients. *Proceedings of the Institution of Mechanical Engineers, 201*(C5), 321–329.

Moore, D. F., & Geyer, W. (1972). A review of adhesion theories for elastomers. *Wear, 22*(2), 113–141.

Myers, N. O. (1962). Characterisation of surface roughness. *Wear, 5*(3), 182–189.

Nakajima, A., Mawatari, T., Yoshida, M., Tani, K., & Nakahira, A. (2000). Effect of coating thickness and slip ratio on durability of thermally sprayed WC cermet coatings in rolling/sliding contact. *Wear, 241*(2), 166–173.

Nowicki, B. (1985). Multiparameter representation of surface roughness. *Wear, 102*(3), 161–176.

Nugent, P. (2008). Quality 101: Surface analysis: Beyond roughness. *Quality Magazine*, April. Available from: http://www.qualitymag.com/articles/85409-quality-101-surface-analysis-beyond-roughness

Poon, C. Y., & Bhushan, B. (1995a). "Comparison of surface roughness measurements by stylus profiler", AFM and non-contact optical profiler. *Wear, 190*(1), 76–88.

Poon, C. Y., & Bhushan, B. (1995b). Surface roughness analysis of glass-ceramic substrates and finished magnetic disks, and Ni-P coated Al–Mg and glass substrates. *Wear, 190*(1), 89–109.

Poon, C. Y., & Bhushan, B. (1996). Nano-asperity contact analysis and surface optimization for magnetic head slider/disk contact. *Wear, 202*(1), 83–98.

Proctor, T. D. (1993). Slipping accidents in Great Britain—An update. *Safety Science, 16*(3–4), 367–377.

Rabinowicz, E. (1965). *Friction and wear of materials*. New York: Wiley.

Radhakrishnan, V. (1970). Effects of stylus radius on the roughness values measured with tracing stylus instruments. *Wear, 16*(5), 325–335.

Roth, F. L., Driscoll, R. L., & Holt, W. L. (1942). Frictional properties of rubber. *Journal of Research of the National Bureau of Standards, 28,* 439–464.

Rowland, F. J., Jones, C., & Manning, D. P. (1996). Surface roughness of footwear soling materials: relevance to slip-resistance. *Journal of Testing and Evaluation, 24*(6), 368–376.

Sanderson, M. J., Smith, I., Parker, I., & Bootman, M. D. (2014). *"Fluorescence microscopy", Cold Spring Harbour Protocols* (pp. 1042–1065). Cold Spring Harbor Laboratory Press.

Schaffer, G. H. (1988). The many faces of surface texture. Special Report 801 (pp. 61–68), American Machinist & Automated Manufacturing, June.

Shawn Janzen, S. (2016). *Descriptive statistics.* Shawn's Statistics Tutoring Pages. Available from: http://www.janzengroup.net/stats/about.html

Sherrington, I., & Smith, E. H. (1987). Parameters for characterizing the surface topography of engineering components. *Proceedings of the Institution of Mechanical Engineers, Part C: Journal of Mechanical Engineering Science, 201*(4), 297–306.

Spragg, R. C., & Whitehouse, D. J. (1970). A new unified approach to surface metrology. *Proceedings of the Institution of Mechanical Engineers, 185*(1), 697–707.

Stachowiak, G. W., & Batchelor, A. W. (2005). *Engineering tribology* (3rd ed.). Saint Louis, Missouri, USA: Butterworth-Heinemann.

Stout, K. J., & Davis, E. J. (1984). Surface topography of cylinder bores—The relationship between manufacture, characterization and function. *Wear, 95*(2), 111–125.

Tabor, D. (1974). Friction, adhesion and boundary lubrication of polymers. In L. H. Lee (Ed.), *Polymer science and technology (5A)* (pp. 54–58). New York: Plenum Press.

Thomas, T. R. (1981). Characterization of surface roughness. *Precision Engineering, 3*(2), 97–104.

Thomas, T. R. (1982). *Rough surfaces.* New York: Longman.

Wilson, M. P. (1996). Slip resistance characteristics of footwear solings assessed using the SATRA friction tester. *Journal of Testing and Evaluation, 24*(6), 377–385.

Wilson, T., & Sheppard, C. (1984). *Theory and practice of scanning optical microscope.* London: Academic Press.

Woo, K. L., & Thomas, T. R. (1979). Roughness, friction and wear: The effect of contact platform. *Wear, 57*(2), 357–363.

Chapter 7
A Practical Design Search for Optimal Floor Surface Finishes—A Case Study

7.1 Introduction

Although many causes are involved in the slip and fall incidents, footpaths and walkways have a major effect on safe ambulation of the pedestrians. Pedestrian floor surfaces should be constructed to provide comfortable walking environments and good traction functioning against any slippery condition. They also should deliver optimal slip resistance qualities throughout their lifetimes. Whilst supporting and controlling slip resistance properties of the floor surfaces would be desirable as a general rule, a specific problem arising in the real world' walking environments is that the slip resistance of floors and floor coverings deteriorates over time periods due to a number of reasons such as material ageing, surface wear, soiling, and maintenance (Leclercq and Saulnier 2002).

Derler et al. (2008) investigated the shift of friction coefficients (COF) for various floor coverings and different test sites over a period of 30 months, in order to study short- and long-term effects of use and maintenance in detail. They reported that mechanical abrasions and coatings by care products led to the continuous reduction of slip resistance properties, a typical outcome for many floor surfaces in use.

These results were also found in the field studies on which the frictional behaviours of floors and floor coverings were investigated by testing a number of different factors. For example, diverse test sites, mechanical erosions, solings, and maintenance and cleaning issues strongly affected the surface texture and the complex interplay of these aspects could lead to considerable local variations of surface features (Chang et al. 2003, 2008; Li et al. 2004).

The recent literature also has stressed the importance of surface coarseness on slip resistance assessments. A number of surface roughness parameters were measured to identify correlations between the surface roughness and slip resistance properties (Kim 1996a, b, c, d, 2004a, b, 2005a, b; 2006a, b, 2015a, b, c, d, 2016a, b;

I.-J. Kim, *Pedestrian Fall Safety Assessments*,
DOI 10.1007/978-3-319-56242-1_7

Kim and Smith 1998a, b, 1999, 2000, 2001a, b, c, 2003; Chang 2001, 2002; Kim et al. 2001, 2013; Kim and Nagata 2008a, b).

Those studies reported that surface roughness on the shoes and floors significantly affected slip resistance performance. Surface roughness offers drainage spaces to avoid squeeze film formations under contaminated environments. For example, tread patterns on the heel surface could improve traction properties by providing void spaces for the removal of lubricants and leading to an increase in direct contact with the floor surface. Therefore, macro-roughness or tread patterns are commonly designed into the shoe heel/sole areas, but they become ineffective quickly after being worn.

However, floor surfaces seem to provide better effects on slip resistance performance than shoes because floor finishes may offer sharper, taller, and tougher asperities in their surface features than shoe ones (Kim et al. 2001; Kim 2004a, b). Although intensifying slip resistance properties of the floor surfaces would be desirable as a general rule, a very high level of traction or slip resistance may impede safe and comfortable ambulation (Chaffin et al. 1992). Moreover, maintaining and/or increasing the surface roughness of the floors and floor coverings require high processing costs.

Despite considerable experimental and analytical research on the slip and falls incidents, no simple concept and/or theoretical model has been developed to explore the effect of floor surface finishes on slip resistance performance. It is also hard to find any definitive studies and design information on operational levels of floor surface roughness required for optimal slip resistance performance. There are no internationally accepted guidelines and design data on operational levels of floor surface finishes for effective controls of slip resistance functioning. Therefore, it is necessary to develop a method which can provide practical information on floor surface finishes for efficient control of slip resistance performance under a range of walking environments.

This chapter aims to investigate the effects of floor surface roughness on slip resistance properties and identify operational levels of floor surface finishes for optimal slip resistance performance under a range of slippery environments. This approach helps readers to develop a concept on design issues of floor surfaces and recognise operational levels of floor surface roughness for best slip resistance performance against a variety of slip hazardous circumstances. To achieve this goal, the present chapter conducted slip resistance tests under three different risky levels of unsafe walking environments such as mildly slippery condition (tap-water covered wet), moderately high slippery one (soapsuds-covered soapy environment), and highly slippery one (machine oil-covered oily) and compared with the dry one.

The main approach was based on the theory concept and model developed by Kim et al. (2013). It is expected that collected information on operative ranges of floor surface roughness under diverse walking environments can be used as a reference to improve floor surface finishes and accordingly a valuable source to develop practical design information and guidelines for floor surfaces required to prevent pedestrian slip and fall incidents.

7.2 Theory Development

7.2.1 Main Hypothesis

Frictional behaviours are the results of extremely complex interactions between the surface and near-surface regions of two materials in repetitive contact and sliding events. Physical, chemical, and mechanical properties of the surface and near-surface regions may well differ from the corresponding bulk properties of parent materials (Kim and Nagata 2008b).

In addition, these surfaces and near-surface regions can change radically as a result of interactions of the surface molecules with their environments throughout the course of repetitive friction processes (Kim and Nagata 2008b). Hence, it is necessary to investigate tribo-physical behaviours between the shoe heel and floor surface and their interactions on slip resistance properties.

Almost all surfaces are rough on a molecular level and variations in the surface profiles can be represented by a random arrangement of peaks and valleys or asperities (Johnson 1985; Kim 2004a, 2006a; Stachowiak and Batchelor 2005; Kim and Nagata 2008a, b). When a shoe heel is loaded against a floor, it can be assumed that two surfaces initially touch only at tiny discrete areas where their highest asperities are in contact, as illustrated in Fig. 7.1a.

A local pressure at the contact region seems to be high enough to cause plastic deformation of the heel asperities even at the lightest load because an elastic

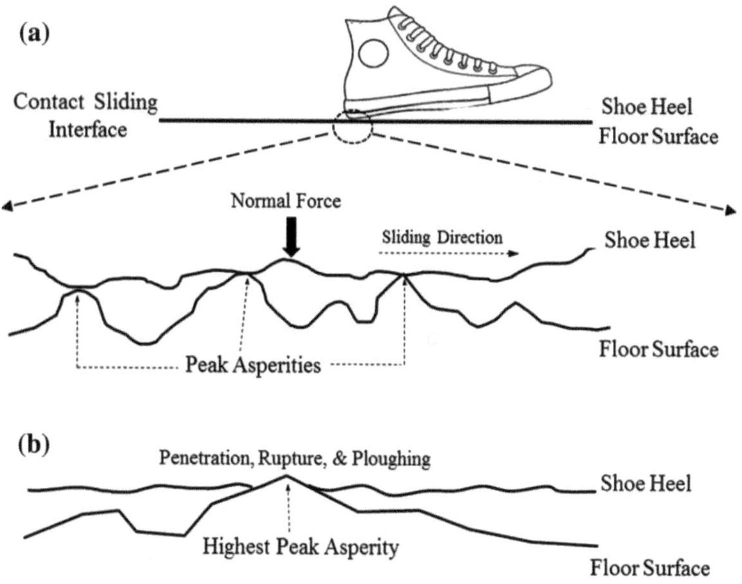

Fig. 7.1 Schematic illustrations for the detailed images of the contact-sliding interface between a shoe heel and a floor surface **a** initial contact state and **b** interlocking mechanism

modulus of the shoe heel is significantly less than that of the floor surface. Thus, its contact mechanism is likely to be an interlocking status as shown in Fig. 7.1b with a magnified format.

If the shoe heel slides on the floor surface, then the heel surface would be ruptured, deformed, and ploughed by wedge-shaped hard asperities of the floor surface. In this model, it can be hypothesised that peak heights of the floor surface are a vital factor affecting the heel contact-sliding mechanism. Therefore, high-peak related roughness parameters of the floor surface and their effects on slip resistance properties are mainly considered in this chapter.

For the high-peak related roughness parameters, $\boldsymbol{R_t}$ (maximum peak-to-valley height) and $\boldsymbol{R_{tm}}$ (maximum mean peak-to-valley height) are considered with $\boldsymbol{R_a}$ (centre line average), which is the most widely used roughness parameter (Thomas 1999; Bewoor and Kulkarni 2009). These three roughness parameters have been commonly used for assessing the effects of floor surface roughness on slip resistance performance (Harris and Shaw 1988; Proctor 1993; Manning and Jones 1994, 2001; Kim and Smith 2000, 2003; Chang et al. 2001; Kim et al. 2001, 2013; Leclercq and Saulnier 2002; Kim 2004a, b, 2015a, b, c; Li et al. 2004; Kim and Nagata 2008a, b).

Detailed information on the three roughness parameters are found in the recent literature (BSI 1988; ISO 1998; Kim and Smith 2000; Chang et al. 2001) and Chap. 6 of this book as well.

7.2.2 A Floor-Surface Model for Optimal Operational Levels

It is hypothesised that floor surface roughness impacts a DFC in a non-linear manner and its effect can be characterised by three operative zones: initial low-growth (Zone 1), mid steady-growth (Zone 2), and top no-growth or peak (Zone 3) as suggested in Fig. 7.2.

In the Zone 1, it can be assumed that multiple mechanisms of frictional behaviours such as adhesion, abrasion, ploughing, and fatigue are involved and the interlocking is ineffective at this stage (Kim 2015b). The Zone 2 supposes to reflect a dominance of the interlocking mechanism where a linear relationship may be valid whilst the Zone 3 indicates an exhaustion of the interlocking mechanism ("no further benefit ").

For practical design purposes, it seems to be beneficial to find the most effective levels ("operational range") of floor surface roughness, which corresponds to the DFC steady-growth zone, as shown (a shadow area) n Fig. 7.2. A lower bound for the operational range can be determined by a safety requirement of DFC > 0.4 (AS/NZS 4663 2004), whilst an upper bound for the operational range can be at surface roughness scales that do not provide further benefit to DFC quantities (Zone 3).

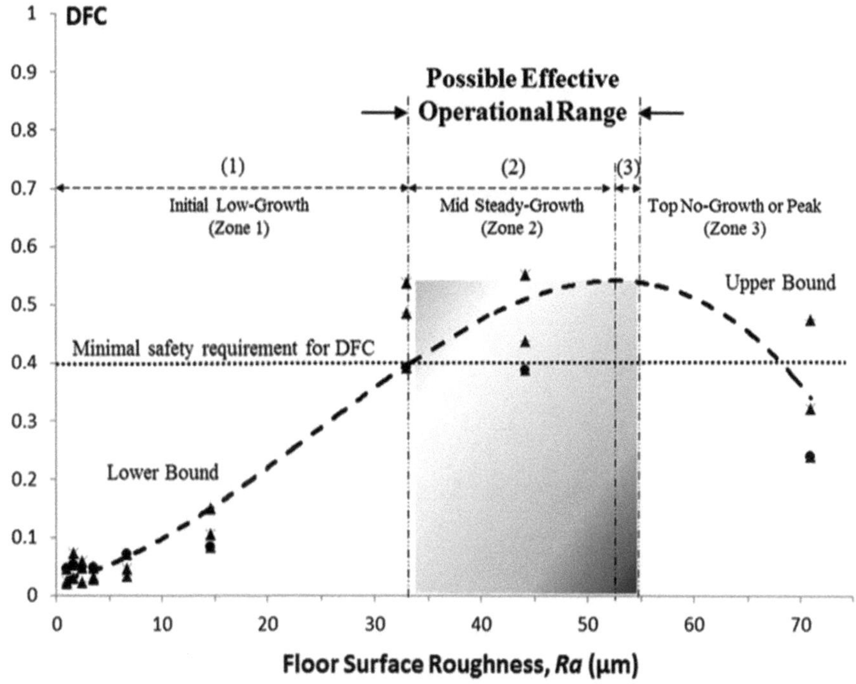

Fig. 7.2 Schematic demonstrations of three operative zones: initial low-growth (Zone 1), mid steady-growth (Zone 2), and top no-growth or peak (Zone 3)

7.3 A Case Study—Experimental Methods and Materials

7.3.1 Dynamic Friction Tester

A pendulum-type dynamic friction tester was used to measure slip resistance properties between a shoe and floor specimen as shown in Fig. 7.3. This test rig consists of two hydraulic systems with an attached artificial foot, a force component transducer (Kistler 3-Component Dynamometer, type 9257A), an angular displacement transducer, and a desktop computer (Kim et al. 2013). This tester simulates movements and loadings of the foot during heel strikes and initial slips and quantitatively determines slip resistance in terms of a dynamic friction coefficient (DFC).

The tester's setup values could be adjusted to cover various parameters taken from human walking trials such as a heel contact angle, vertical load, and its rate of increase and sliding speed (Kim et al. 2013). To adjust the heel contact angle, tapered shims were inserted between the pendulum base and the last on which the shoe was mounted.

The vertical load was set by the pendulum length adjusted by two nuts. Some fine adjustment to the vertical load was also possible by a pressure control on the oil

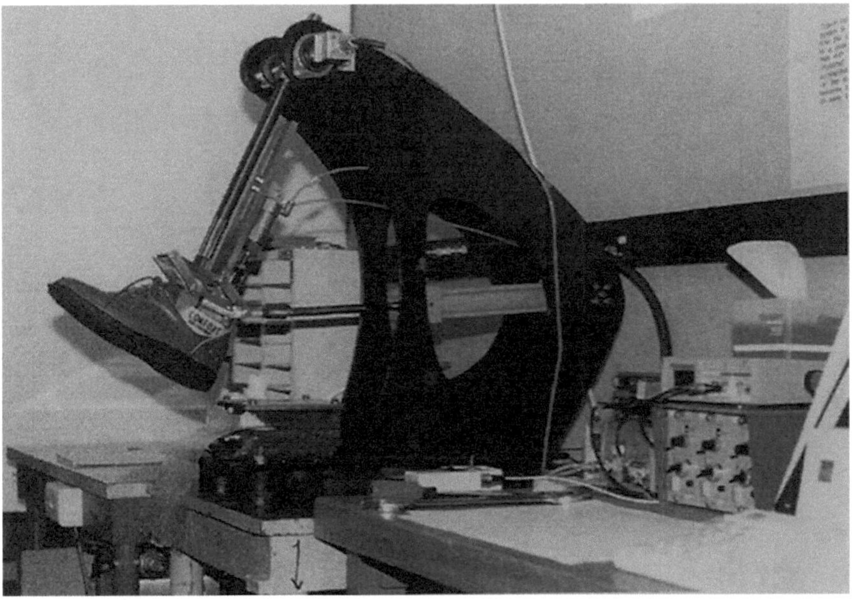

Fig. 7.3 A photographic image of dynamic friction measuring device

fed to the vertical hydraulic cylinder. The sliding speed was set by a flow control valve. In this tester, the test shoe was firmly attached to the last and mounted at the end of the pendulum mechanism. In order to minimise any movement, the shoe specimen was nailed to the pendulum last.

The floor specimen was glued to a steel plate that was bolted onto the force component transducer. In each test, the shoe was driven forward by the horizontal hydraulic cylinder to contact the floor surface at the heel edge. Another hydraulic cylinder was mounted at the pendulum end to simulate the body weight portion supported by the leading foot at heel strike. The two hydraulic cylinders were in a common circuit supplied by a pump that was driven by an electric motor.

As the shoe heel passed across the floor surface, the frictional (horizontal: *H*) and normal (vertical: *V*) components of the resultant force were measured by a Kistler dynamometer on which the floor sample was firmly mounted. The test speed was measured by a rotary potentiometer driven by the pendulum shaft.

This instrument produced two separate signals that were proportional to the frictional (horizontal) and normal (vertical) force components, respectively. The two hydraulic cylinders were in a common circuit supplied by a pump which was driven by an electric motor. The force component signals and potentiometer voltages during a test were recorded on a personal computer which continually calculated the *H/V* force ratio. The force components, their ratio and the angular position were then drawn on a computer screen (see Fig. 7.4).

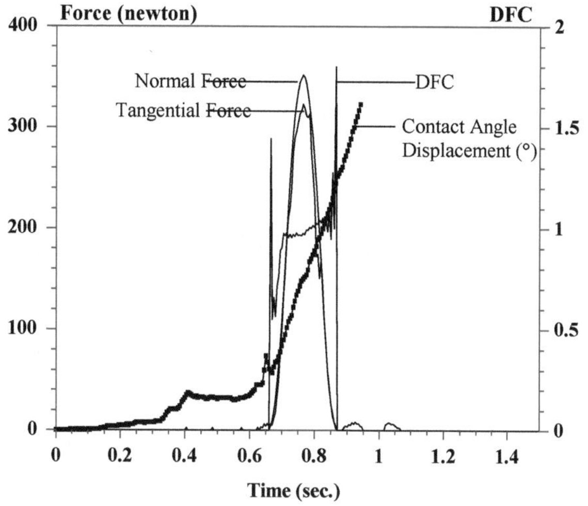

Fig. 7.4 A typical output from the pendulum type dynamic friction tester was recorded by a desktop computer that continually calculated an *H/V* force ratio

7.3.2 *Test System Conditions*

7.3.2.1 Test Speed

It seems more controversial to choose the appropriate slipping speed at which to drive the shoe heel across the floor specimen. Normal walking speed varies from 1 to 2 m/s, but the forward speed of shoe heel/sole edge seems to be probably less than this speed just before heel strike.

Strandberg (1983) showed experimental results for different subjects, in which walking speed varied from 0.06 to 0.32 m/s. After sliding started, the shoe heel accelerated to a value above that of walking range between 0 and 0.5 m/s.

Perkins and Wilson (1983) also concluded from their experimental results that "probably the ideal speed for a high-speed measurement is 0.5–1.0 m/s since the foot and shoe can be travelling at this speed when the shoe heel tip touches the floor surface. Even if slip starts from a static situation, such is the acceleration that the speed of slip is about 0.5 m/s after only 0.01 m slip distance".

In a series of tests performed by Hoang et al. (1985, 1987), they found out that there were good results with speed variations (less than 8%) within the speed range 0.2–0.6 m/s. On the basis of Hoang et al.'s measurement results, the test speed for all the slip resistance measurement tests in this book was kept at a constant speed of 0.4 m/s.

7.3.2.2 Contact Angle

The strike angle at which the shoe heel first contacts the floor surface is an important factor in slipping. This contact angle has proved to be an important variable because of its influence on the nominal contact area between the shoe heel and floor surface. In fact, the contact area plays a dominant role in the slipping processes when walking on lubricated surface conditions (Moore 1972, 1975). Therefore, observing the contact angle at a "correct angle" would mean testing one of the most important parts of the slip resistance properties of shoe heels/soles in walking.

The contact angle to be used was selected after a number of biomechanical studies of the foot angles of people walking on the force-measuring platform. Thirty-two male subjects ranging from 155 to 180 cm height were selected and tested. Experimental results from the biomechanical trials with human subjects showed that the contact angle between the shoe heel and floor surface varied with the walking speed, step size and person's height (Hoang et al. 1987).

On a horizontal surface, it laid within the range of 6–10° measured from the floor surface. Analysing the distribution, an average heel strike angle of 9° was chosen for the dynamic friction tester. The of shoe heel can be adjusted by tapered shims between the shoe last and the vertical pendulum.

7.3.2.3 Normal Force

The hydraulic system for the normal force in the friction tester can provide a constant slipping velocity at up to 500 N of vertical loads (Hoang and Stevenson 1981). According to Strandberg and Lanshammar (1981), they uncovered that the leading foot carried up to 60% of body weight, acting at the heel rear edge, during the shoe heels contacted to the floor surfaces. Therefore, a maximum vertical component of the resultant force was selected close to 350 N, which is an about half of the weight of an average person, and kept constant throughout the tests.

7.3.2.4 Dynamic Friction Test Output

An output of a typical dynamic friction test is shown in Fig. 7.4. The dynamic coefficient of friction (DCOF or DFC) in a particular test was measured from the central section of the recorded trace as shown in Fig. 7.4, where the value was fairly steady. In most cases, there was an initial peak (beginning part of the plot) on the slip resistance trace, and in some cases, this peak was most pronounced and repeated in a certain amount of test intervals. Some peaks were also observed on the rear side on the slip resistance trace (last part of the plot). It is considered that these peaks are likely to relate to topographic changes of the sliding interface between the shoe heel and floor surface. Since the heel surface may suffer a massive amount of pressure caused by an initial contact stage and geometric peel-off caused by followed sliding friction events. That is, the shoe surface seems to experience wear developments from repetitive abrasive processes.

The DFC was calculated by dividing the horizontal force component with the vertical force one. Its output and the associated angular position of the pendulum were collected by a personal computer.

During the dynamic friction tests in this book, the normal force was maintained around 350 N and the sliding speed was controlled at 0.4 m/s based on gait studies (Kim et al. 2013). A heel contact angle of 9° was also chosen from the result of previous biomechanical studies (Kim et al. 2013).

7.3.3 Floor and Shoe Specimens

For floor specimens, nine commercially available new flooring materials were used for the dynamic friction tests. The floor specimens were carefully chosen to include surfaces within a wide range of surface roughness. Table 7.1 summarises the floor specimens for the dynamic friction tests.

For shoe specimens, three commercially available new shoes were used for the tests. They included two Nitrile Rubber (S1: Nitrile Rubber No. 1 and S2: Nitrile Rubber No. 2) and a PVC (S3) soles and heels, respectively.

The floor and shoe specimens were thoroughly cleaned with demineralized water to eliminate any dirt and dust, and dried and kept in plastic containers during the tests.

Table 7.1 Summary of the floor specimens with surface roughness parameters—R_a, R_t, and R_{tm}

Floor specimen name	Surface roughness parameter (μm)		
	R_a^a	R_t^b	R_{tm}^c
Terrazzo	0.96	8.23	4.85
Smooth vinyl tile	1.55	13.61	10.26
Smooth metal plate	2.36	13.38	11.76
Smooth ceramic tile	3.43	27.50	17.29
Smooth concrete slab	6.59	54.00	35.80
Moderate rough ceramic tile	14.54	85.51	61.75
Moderate rough concrete slab	32.97	337.00	224.33
Rough concrete slab	44.11	226.75	159.25
Rough ceramic tile	70.94	396.80	141.00

[a] R_a = Center line average
[b] R_t = Maximum peak-to-valley height
[c] R_{tm} = Maximum mean peak-to-valley height

7.3.4 Environmental Conditions

Dynamic friction tests were conducted under four different environments: (1) clean and dry, (2) tap water-covered wet, (3) soapsuds-covered soapy and (4) machine oil-covered oily conditions. A commercial type detergent (Kinematic Viscosity: 1.27 *cSt* at 16 °C) and machine oil (Kinematic Viscosity: 343 *cSt* at 16 °C) were applied to create soapy and oily environmental conditions, respectively.

A fixed amount (15 ml) of tap water, a mixture of water and detergents, and machine oils were sprayed over the whole floor surface (specimen size: 110 × 170 mm) to create different risk levels of slippery environments before the tests. Any debris from the shoe heel was immediately removed using a clean, soft, and dry brush. These procedures were in accordance with the standard parts of BS EN ISO 13287:2012 (International Standard: Personal protective equipment. Footwear test method for slip resistance).

7.3.5 Floor Surface Roughness Measurements

A Talysurf 5 profilometer (Taylor-Hobson, UK) that had a conical stylus with a spherical tip of 12 μm radius was used to measure surface roughness of the floor specimens. To remove waviness components of the floor surfaces, a Gaussian filter was used with a 0.8 mm cutoff over a single traverse length of 17.5 mm.

Surface profiles of each floor specimen were measured five times at three different locations. Measurement results of the surface roughness parameters for each floor specimen were summarised in Table 7.1.

7.3.6 Statistical Analysis and Design

Three-way analysis of variance (ANOVA) was performed to determine the significant effects of floor, shoe and environment variables and their interactions on the DFCs.

Polynomial regression models were used to evaluate the relationships between the floor surface roughness parameters and DFCs under the three different risky levels of unsafe walking environments: mildly slippery condition (tap-water covered wet), moderately high slippery one (soapsuds-covered soapy environment), and highly slippery one (machine oil-covered oily).

Independent variables for the ANOVA included the following items:

(1) the shoe type (“Shoe”) classified by sole/heel materials including two Nitrile Rubber (S1 and S2) and one PVC (S3) heels/soles;

(2) the floor type ("Floor") with nine different coarseness defined by surface roughness scales (measured in $\boldsymbol{R_a}$, $\boldsymbol{R_t}$, and $\boldsymbol{R_{tm}}$); and
(3) the walking environment ("Environment") with three different conditions: tap water-covered wet surface, soapsuds-covered soapy surfaces, and machine oil-covered situations, respectively.

The dependent variable for the ANOVA was a dynamic friction coefficient (DFC). A DFC value of 0.4 was used as an acceptable safety criterion to determine the lower bound of the operational range of floor surface roughness (AS/NZS 4663 2004).

Prevalence odds ratios with a 95% confidence interval were calculated as a measure of association. *P*-values less than 0.05 were considered statistically significant. All statistical analyses were performed using Statistical Analysis System (SAS) software.

7.4 Results of the Case Study

7.4.1 *Slip Resistance Performance*

Figure 7.5 shows the results of dynamic friction tests between the nine floor and three shoe specimens under the four different risk levels of walking environments: dry, wet, soapy, and oily, respectively. The DFC results were arranged by the order of floor surface roughness parameter, $\boldsymbol{R_a}$. This arrangement was intended to analyse the relationship between the surface roughness of each floor specimen and DFC results.

The DFCs generally increased with the surface roughness of floor specimens. This trend was clearly found under the soapy and oily environments than the wet one. However, a linear relationship between the floor surface roughness and DFCs was not found in the cases of dry and wet surfaces. Despite their low scales of surface roughness, some smooth floors (≤ 10 μm in *Ra* roughness parameter) such as the smooth vinyl tile, smooth metal plate, and smooth concrete slab showed good slip resistance performance (DFC $\geq$ 0.4) against all the shoes (except S2) under the wet environment.

7.4.2 *Interactions Between Floor Types and Environments*

The three-way (Floor × Shoe × Environment) ANOVA results in Table 7.2 demonstrates strong interactions amongst the Floor (surface roughness), Shoe, and Environment variables on the DFCs. The DFCs were increased significantly by the floor types with higher surface roughness scales and drastically reduced under the polluted environments. However, the effect of floor type was more significant under the soapy and oily environments than the wet one.

Fig. 7.5 DFC results among the nine-floor surfaces and three shoes under the: **a** clean and dry, **b** tap water-covered wet, **c** soapsuds-covered soapy, and **d** machine oil-covered oily environments, respectively

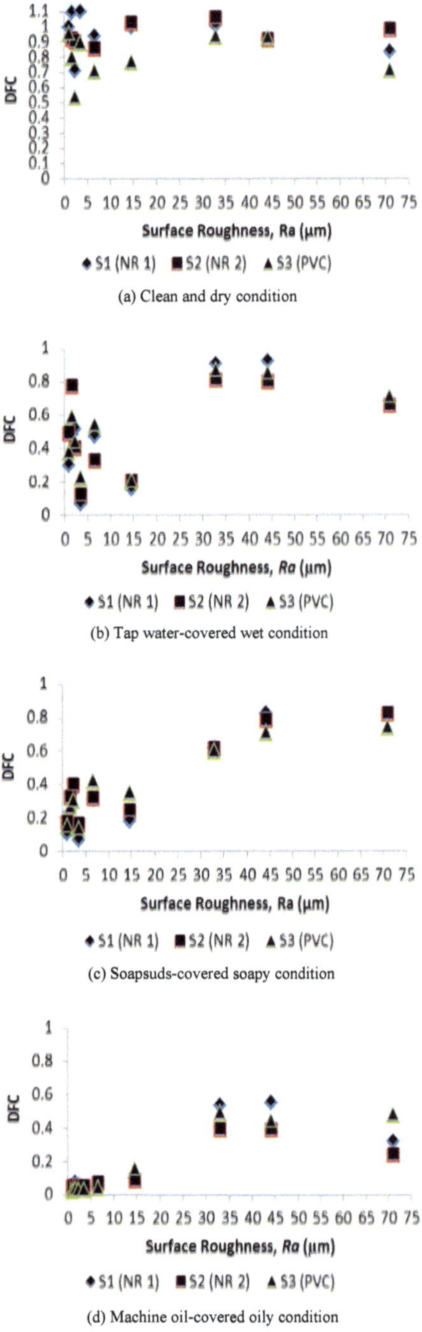

Table 7.2 Summary of three-way analysis of variance (ANOVA) results amongst the shoes, floors, and environments on the DFCs under the wet, soapy, and oily conditions, respectively

Effect on DFC	DF	Sum of squares	Mean square	F value	*Pr* > F
Intercept	1	10.80729	10.80729	3727.37	<0.0001
Floor type[a]	8	3.69161	0.46145	159.15	<0.0001
Environmental condition	2	1.58079	0.79039	272.6	<0.0001
Shoe type	2	0.00298	0.00149	0.51	0.6028
Floor type × Environmental condition	16	0.37556	0.02347	8.1	<0.0001
Floor type × Shoe type	16	0.11600	0.00725	2.5	0.0133
Environmental condition × Shoe type	4	0.01615	0.00404	1.39	0.2588
Error	32	0.09278	0.00290		

[a]Floor type as a categorical variable

Furthermore, as shown in Table 7.3, the DFCs for the environmental conditions were predicted by cubic functions of the three roughness parameters: $\boldsymbol{R_a}$, $\boldsymbol{R_t}$, and $\boldsymbol{R_{tm}}$. They showed fairly good correlations between the surface roughness parameters of floor specimens and DFC results against each shoe type under the three contaminated environmental conditions.

The regression models supported the theory hypothesis and facilitated to derive and visualise the functional levels of floor surface roughness. Tables 7.4, 7.5 and 7.6 summarise the results of detailed regression analyses on the functional parameters against the wet, soapy, and oily environments, respectively. They clearly demonstrate the significance of the three roughness parameters on the DFCs.

The cubic functions of each roughness parameter show that surface roughness has an effect on the DFCs in a non-linear fashion and their properties can be characterised by the three operational zones: Zone 1 (initial low-growth), Zone 2 (mid steady-growth), and Zone 3 (top no-growth or peak), respectively.

Table 7.3 SAS regression procedure: DFCs were predicted by cubic functions of $\boldsymbol{R_a}$, $\boldsymbol{R_t}$, and $\boldsymbol{R_{tm}}$ parameters under the wet, soapy, and oily environments, respectively

Environment	Model	DF	R-square	F value	Pr > F
Wet	$\boldsymbol{R_a}$	3	0.6428	13.80	<0.0001
	$\boldsymbol{R_t}$	3	0.6936	17.35	<0.0001
	$\boldsymbol{R_{tm}}$	3	0.6995	17.84	<0.0001
	$\boldsymbol{R_a}$	3	0.8372	39.43	<0.0001
Soapy	$\boldsymbol{R_t}$	3	0.8738	53.07	<0.0001
	$\boldsymbol{R_{tm}}$	3	0.8665	49.77	<0.0001
	$\boldsymbol{R_a}$	3	0.9029	71.31	<0.0001
Oily	$\boldsymbol{R_t}$	3	0.9309	103.25	<0.0001
	$\boldsymbol{R_{tm}}$	3	0.9295	101.07	<0.0001

Table 7.4 Detailed SAS regression analysis results: DFCs were predicted by cubic functions of R_a, R_t, and R_{tm} roughness parameters under the wet environment

Surface condition	Roughness parameter	Variable	DF	Par. estimate	Standard error	t value	Pr > |t|
Wet	R_a	Intercept	1	0.2188	0.0413	7.45	<0.0001*
		Linear	1	0.0026	0.0095	−2.34	0.0282*
		Quadratic	1	0.0006	0.0004	3.38	0.0026*
		Cubic	1	−8.1E-06	3.31E-06	−3.70	0.0012*
	R_t	Intercept	1	0.2358	0.0397	7.80	<0.0001*
		Linear	1	−0.0008	0.0014	−2.92	0.0078*
		Quadratic	1	2.54E-05	8.73E-06	4.01	0.0006*
		Cubic	1	−5.32E-08	1.44E-08	−4.23	0.0003*
	R_{tm}	Intercept	1	0.2381	0.0445	7.65	<0.0001*
		Linear	1	−0.0014	0.0026	−3.03	0.0060*
		Quadratic	1	5.32E-05	2.72E-05	3.58	0.0016*
		Cubic	1	−1.60E-07	7.66E-08	−3.42	0.0024*

*Indicates significant

Table 7.5 Detailed SAS regression analysis results: DFCs were predicted by cubic functions of R_a, R_t, and R_{tm} roughness parameters under the soapy environment

Surface condition	Roughness parameter	Variable	DF	Par. estimate	Standard error	t value	Pr > |t|
Soapy	R_a	Intercept	1	0.0251	0.0242	5.30	<0.0001*
		Linear	1	0.0036	0.0056	0.28	0.7838
		Quadratic	1	0.0004	0.0002	1.72	0.0990
		Cubic	1	−5.8E-06	1.94E-06	−2.45	0.0225*
	R_t	Intercept	1	0.0379	0.0223	5.94	<0.0001*
		Linear	1	−0.0004	0.0008	−0.58	0.5684
		Quadratic	1	1.81E-05	4.91E-06	2.91	0.0079*
		Cubic	1	−3.85E-08	8.09E-09	−3.69	0.0012*
	R_{tm}	Intercept	1	0.0493	0.0246	5.35	<0.0001*
		Linear	1	−0.0014	0.0014	−0.54	0.5940
		Quadratic	1	4.61E-05	1.5E-05	1.95	0.0630
		Cubic	1	−1.40E-07	4.23E-08	−2.09	0.0479*

*Indicates significant

7.4.3 *Interactions Between Shoe Types and Environments*

The three-way ANOVA results in Table 7.2 demonstrated that there were no interactions between the shoe type (Shoe) and walking environment (Environment) variables on the DFCs. Some of the shoe types performed better than others, but this effect was evident only under the wet surface condition. For example, the shoe

Table 7.6 Detailed SAS regression analysis results: DFCs were predicted by cubic functions of R_a, R_t, and R_{tm} roughness parameters under the oily environment

Surface condition	Roughness parameter	Variable	DF	Par. estimate	Standard error	t value	Pr > |t|
Oily	R_a	Intercept	1	0.2188	0.0413	1.04	0.3109
		Linear	1	0.0026	0.0095	0.65	0.5240
		Quadratic	1	0.0006	0.0004	2.02	0.0552
		Cubic	1	−8.1E-06	3.31E-06	−2.99	0.0066*
	R_t	Intercept	1	0.2358	0.0397	1.70	0.1023
		Linear	1	−0.0008	0.0014	−0.45	0.6582
		Quadratic	1	2.54E-05	8.73E-06	3.68	0.0012*
		Cubic	1	−5.32E-08	1.44E-08	−4.76	<0.0001*
	R_{tm}	Intercept	1	0.2381	0.0445	2.01	0.0564
		Linear	1	−0.0014	0.0026	−0.99	0.3346
		Quadratic	1	5.32E-05	2.72E-05	3.07	0.0054*
		Cubic	1	−1.60E-07	7.66E-08	−3.31	0.0030*

*Indicates significant

S1 was better than the shoes S2 and S3 against certain types of floor surfaces whilst the shoe S3 was better than the shoe S2 generally.

Under the soapy and oily environments, however, all the three shoes were not effective to support slip resistance performance as compared to the floor-type effects. On the other hand, the shoes S1 and S2 were better than the shoe S3 under the clean and dry environmental condition. However, overall effects of the shoe-type were relatively small as compared to the surface roughness related ones.

7.4.4 Operational Ranges of Floor Surface Roughness

Most of the floor types showed high slip resistance performance (DFC > 0.4) under the wet surface condition as shown in Fig. 7.5. But, the DFCs were largely decreased and recorded dangerously low levels (DFC < 0.4) under the soapy and oily environments. Acceptable slip resistance performance under such highly slippery environments was validated only by the three floor specimens with surface roughness >30 μm in R_a roughness parameter: the moderate rough concrete slab (R_a = 32.97 μm), rough concrete slab (R_a = 44.11 μm), and rough ceramic tile (R_a = 70.94 μm), respectively. As summarised in Table 7.1, R_t and R_{tm} parameters of these three-floor specimens also showed correspondingly high scales. However, the rougher floors such as the rough concrete slab and rough ceramic tile did not provide any further increase of DFCs under the three lubricated environments as compared to the moderate rough concrete slab.

Figures 7.6, 7.7, and 7.8 show the polynomial regression results for R_a, R_t, and R_{tm} roughness parameters with regression lines of the DFCs under the three

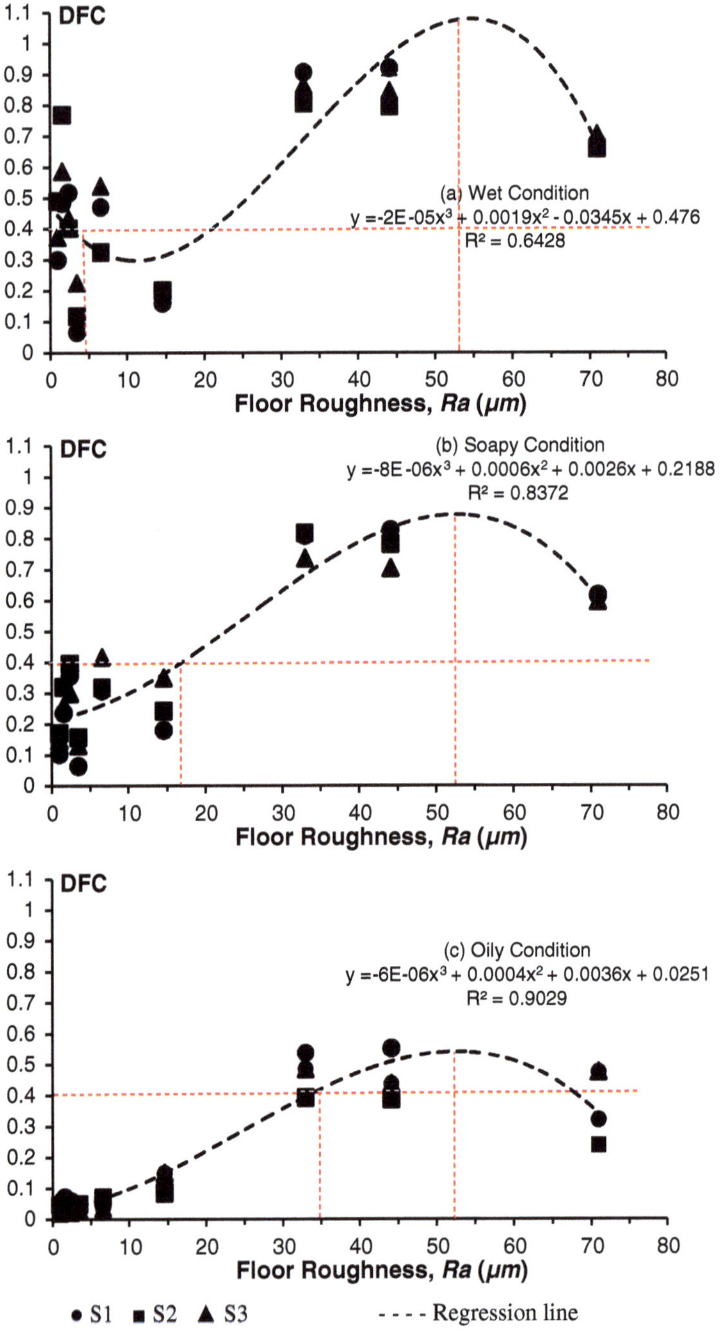

Fig. 7.6 Scattered plots and polynomial regression lines of the DFCs and the floor surface roughness parameter, R_a under the **a** wet, **b** soapy and **c** oily conditions, respectively

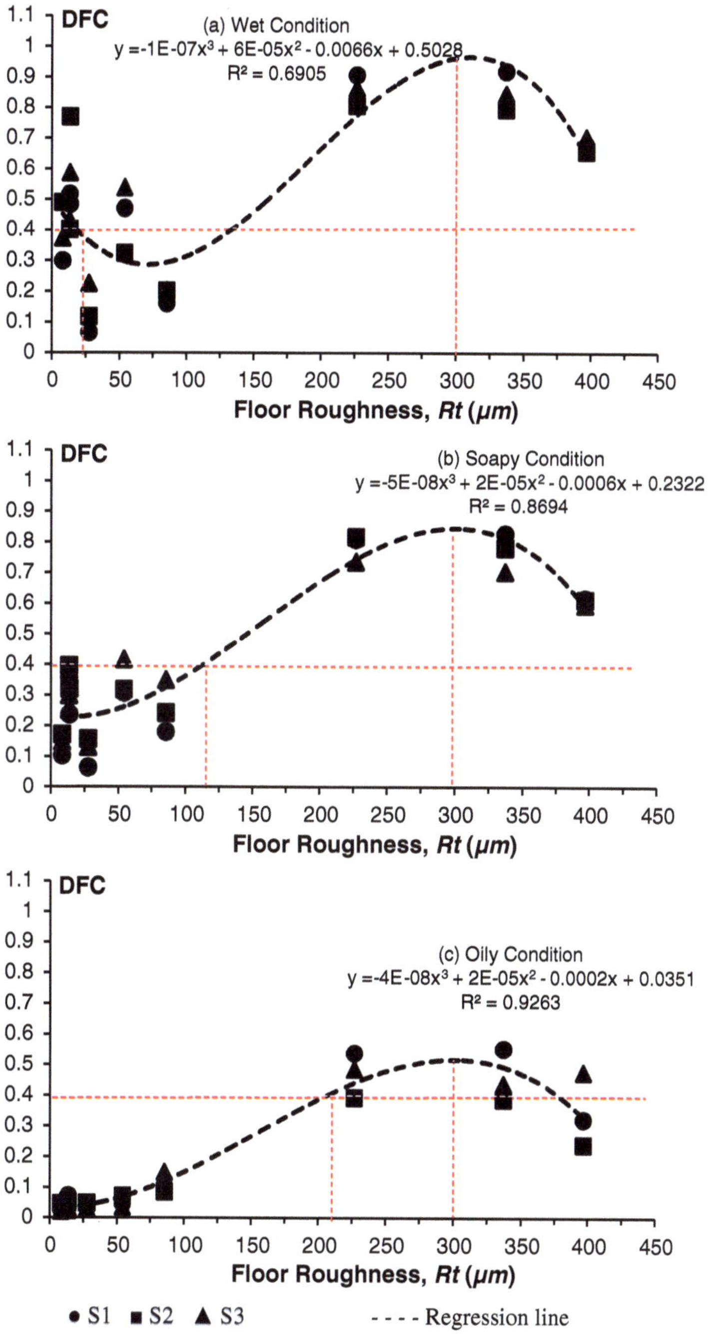

Fig. 7.7 Scattered plots and polynomial regression lines of the DFCs and the floor surface roughness parameter, R_t under the **a** wet, **b** soapy and **c** oily conditions, respectively

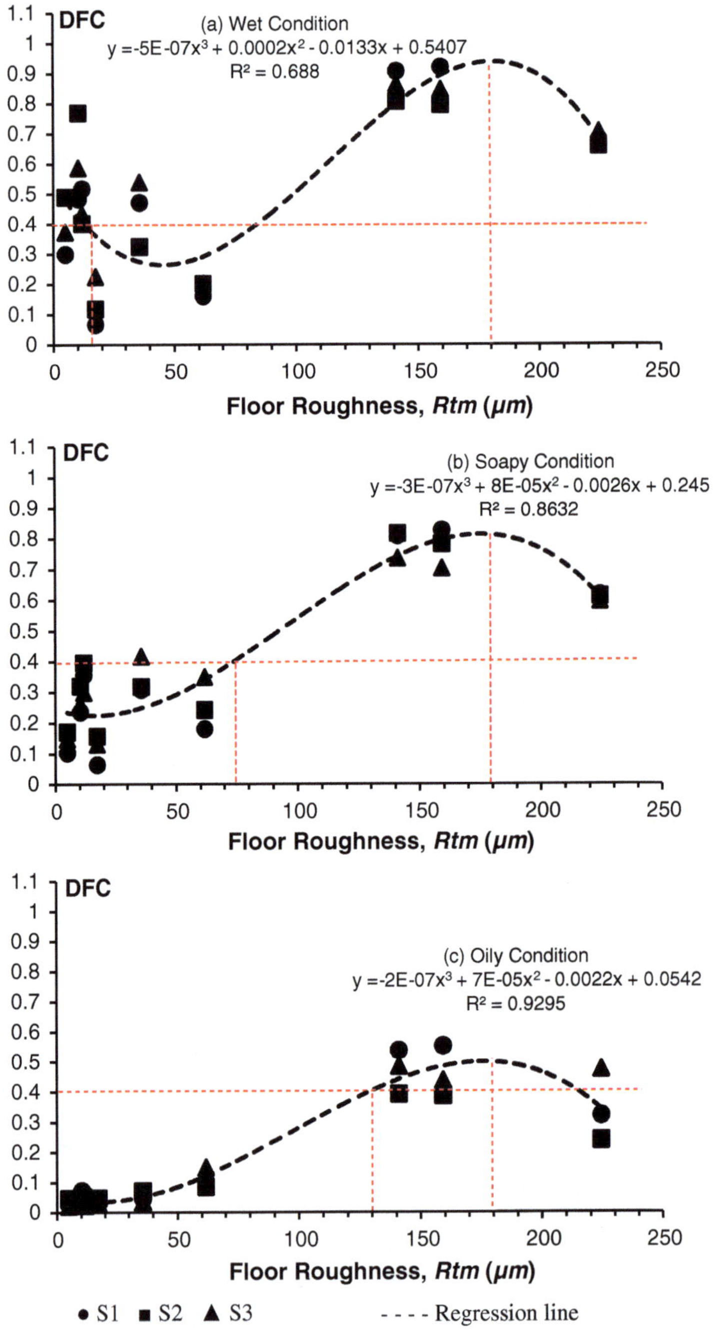

Fig. 7.8 Scattered plots and polynomial regression lines of the DFCs and the floor surface roughness parameter, $\boldsymbol{R_{tm}}$ under the **a** wet, **b** soapy and **c** oily conditions, respectively

lubricated conditions. The cubic functions of regression equations demonstrate examples of possible operational relationships between the floor surface roughness parameters and DFCs against each environmental condition. Using the regression models and the safety requirement for DFC > 0.4, it could be estimated that the lower bound ranges for floor surface roughness would be at R_a = 35 μm, R_t = 210 μm, and R_{tm} = 130 μm for the oily environment and at R_a = 17 μm, R_t = 120 μm, and R_{tm} = 75 μm for the soapy environment, respectively.

The regression curves in Figs. 7.6, 7.7, and 7.8 also present that with increments of floor surface roughness the DFCs reach a top no-growth or a peak (Zone 3) after certain scales of floor surface roughness under the soapy and oily environments. From the regression models, it could be predicted that the upper bound ranges for floor surface roughness would be at R_a = 52 μm, R_t = 300 μm, and R_{tm} = 180 μm for the oily condition, and at R_a = 52 μm, R_t = 300 μm, and R_{tm} = 180 μm for the soapy condition, respectively. Those results could be considered as lower and upper bounds for the operational ranges of floor surface roughness under both polluted environments.

Some of the smooth floor surfaces such as the smooth vinyl tile (R_a = 1.55 μm), smooth metal plate (R_a = 2.36 μm), and smooth concrete slab (R_a = 6.59 μm) showed relatively good slip resistance performance under the wet and soapy surface conditions. But, their slip resistance properties were dangerously low against most of the soapy and oily conditions thus should not be treated as an exception from the suggested operational ranges.

7.5 Assessments and Verifications of Findings

7.5.1 Interactions Between Floor Types and Environments

There were significant correlations between the floor surface roughness and DFCs under the three lubricated environments. When the floor surfaces were polluted, the role of floor surface finishes was very significant to increase slip resistance performance. This effect was noticeable under the soapy and oily conditions.

As shown in Fig. 7.5, there were large increases in the DFCs as the surface roughness was increased from 14.54 to 32.97 μm in R_a roughness parameter. This increment was even sufficient enough to raise slip resistance performance from a dangerously low level to marginally safe one in the DFCs (>0.4) under the oily environment. However, further increases of the floor surface roughness did not provide additional benefits to the slip resistance performance against all the three environments. This result was consistent with other studies reported on functional relationships between the surface roughness and DFCs under contaminated conditions (Manning and Jones 2001; Chang 2001, 2002; Kim et al. 2013; Kim 2015b)

It was also found that there was a lack of correlations between the floor types and DFCs under the dry and wet environments. This finding indicates the involvement of multiple mechanisms of slip resistance properties and possible effects of other tribo-physical characteristics such as wear developments and hydrodynamic and boundary lubrication effects amongst the floor surfaces, shoe heels, and environments during the dynamic friction tests. This aspect requires further studies to understand their complex mechanisms and effects on slip resistance performance.

7.5.2 Interactions Between Shoe Types and Environments

The interactions between the shoe types and environmental conditions demonstrated that there were no shoe effects to improve slip resistance performance under the soapy and oily environments (see, Table 7.2). This result suggests that introducing rough floor surfaces would be a more effective strategy than shoes to increase slip resistance performance under such highly slippery environments.

7.5.3 Operational Ranges of Floor Surface Roughness

The test results showed non-linear relationships between the floor surface roughness and DFCs. This finding confirmed that the higher slip resistance performance was not mechanically supported by the rougher floors as assumed in the earlier study (Kim et al. 2013).

Some smooth floors such as the smooth vinyl tile (1.55 μm in $\boldsymbol{R_a}$ roughness parameter) and smooth metal plate (2.36 μm in $\boldsymbol{R_a}$ roughness parameter) showed better DFC results than the rougher ones (>30 μm in $\boldsymbol{R_a}$ roughness parameter) under the wet condition (see, Fig. 7.5b). Because of this aspect of slip resistance properties, it was difficult to estimate the lower bound ranges for floor surface roughness under the wet surface condition.

The test results also identified that increasing the floor surface roughness beyond certain scales did not provide further benefits for slip resistance performance. This means that there seems to exist powerful ranges of operational roughness for optimal slip resistance performance.

For example, the floor surface roughness in the scales of 35–52 μm ($\boldsymbol{R_a}$ roughness parameter) seems to be an efficient range for the oily environment whilst the floor surface roughness in the scale of 17–52 μm ($\boldsymbol{R_a}$ roughness parameter) would be an effective one for the soapy environment, respectively.

This outcome demonstrates that the oily environment requires twice rougher floor surface roughness than the soapy one in their lower boundary roughness scales: 35 μm versus 17 μm in $\boldsymbol{R_a}$ roughness parameter. On the other hand, it is interesting to note that the upper bound of floor surface roughness shows the same ranges of surface roughness scales: 52 μm in $\boldsymbol{R_a}$, 300 μm in $\boldsymbol{R_t}$, and 180 μm in $\boldsymbol{R_{tm}}$ roughness parameters under the three lubricated environments.

Table 7.7 Summary of operational ranges with the lower and upper bounds of the floor surface roughness parameters for optimal slip resistance performance under the three lubricated environments

Surface roughness parameters	Operational ranges (μm) with lower and upper bounds under three lubricated environments		
	Wet	Soapy	Oily
R_a	5 (21)–52	17–52	35–52
R_t	25 (135)–300	120–300	210–300
R_{tm}	15 (80)–180	75–180	130–180

Table 7.7 summarises operational ranges with the lower and upper bounds of the floor surface roughness parameters for optimal slip resistance performance against the three polluted environments.

As discussed by Kim et al. (2013), there was no study found a concept of the upper bound of floor surface roughness and its operational ranges for effective slip resistance controls. The present chapter adopted the criteria of "no further benefit for DFC values" to determine the upper boundary of floor surface roughness for slip resistance performance (Kim et al. 2013; Kim 2015b). Other cut-off criteria to determine the upper boundary of floor surface roughness could be derived from the requirement of safe and comfortable ambulation to avoid falling, but the associated critical DFC values that could be used for such criteria have not been recognised yet.

The other issue to be considered would be dealing with developments of wear and tear process of floor surfaces with their long-term usages (Kim 2015b). With sustained ambulation, the floor finishes will be continuously modified by wear progression and accordingly operational ranges of floor surfaces will be modified from optimal ones. This feature requires further studies to understand wear behaviours and their effects on slip resistance performance.

However, findings on the operational ranges of floor surface roughness signify that we do not always need to choose rougher floors and/or treat floor surfaces to roughening, which may have important applications to the practical design developments for floors and floor coverings as reported in the recent studies (Kim et al. 2013; Kim 2015b). Hence, floor surfaces seem to require different levels of coarseness for different types of environmental conditions to effectively and efficiently manage walkway slipperiness.

7.6 Study Limitations

This study tested a small selection of floor types and floor materials with restricted surface roughness scales. Further research on the slip resistance properties of flooring materials with different surface finishes is required to determine the specific functional levels of surface roughness.

The test design was also constrained by three polluted environmental conditions. Those experimental settings may limit the applicability of findings only to those types of polluted environments. Other types of contaminants and environmental situations with different compositions and viscosities may result in different functional levels of floor surface roughness.

Further research also requires to collecting data for specific operational ranges of floor surface roughness with different material types and heel/sole designs of shoes against a range of walking circumstances.

7.7 Conclusions

This chapter aimed to investigate the effects of floor surface finishes on slip resistance properties and identify operational ranges of floor surface roughness for optimal slip resistance performance under different slippery environments. The study results showed that all the tested floor-shoe combinations provided insufficient slip resistance performance under the lubricated environments. However, there was a lack of correlations between the surface roughness and DFCs under the dry and wet surface conditions. This finding indicated the involvement of complex mechanisms of slip resistance properties and possible effects of other tribo-physical characteristics amongst the floor surfaces, shoe heels, and environments on the slip resistance performance.

Slip resistance performance was significantly affected by and well correlated with floor surface finishes under the highly slippery environments such as soapy and oily conditions. Polynomial regression models for the floor surface roughness and DFC interactions allowed estimating operational ranges of floor surface roughness for optimal slip resistance performance. Floor surfaces with around 17–52 μm and 35–52 μm in $\boldsymbol{R_a}$ roughness parameter most likely represented the lower and upper bounds of operational ranges for the best slip resistance controls under the soapy and oily environments, respectively.

Overall results evidently demonstrate that the proposed concept on operational ranges with the lower and upper bounds for the floor surface roughness may have applicable design implications for floors and floor coverings to reduce slip and fall hazards. The case study explored in this chapter suggests that designs for the floor surface finishes would require different levels of surface roughness for different types of environmental conditions to effectively control slippery walking environments.

7.8 Chapter Summary

Increasing traction properties of the floor surface would be desirable as a general rule, but a very high level of slip resistance may impede safe and comfortable ambulation. There is a lack of evidence whether traction properties are linearly correlated with surface features of the floor or what levels of floor surface finishes are required to effective control of slipperiness. It is also scarce to find studies and/or guidelines on the operational ranges of floor surface roughness required for optimal slip resistance performance.

The main objectives of this chapter are to investigate the effects of floor surface finishes on slip resistance performance under different environmental and shoe-type conditions and identify operational ranges of floor surface roughness as practical design information for the effective control of fall incidents.

A theory model of three operative zones was suggested to characterise functional levels of floor surface roughness on slip resistance performance. To test the theory model, dynamic friction tests were conducted using 3 shoes and 9 floor specimens under 4 different environments: clean and dry, wet, soapy, and oily conditions. The test results showed that significant effects of floor-type on DFCs were found in the polluted environments. As compared to the floor-type effect, the shoe-type effect was relatively small. Slip resistance performance was significantly affected by and well correlated with the floor surface roughness under the soapy and oily environments. Polynomial regression analyses amongst the floor surface roughness, DFCs and environments allowed to estimating operational ranges for optimal slip resistance performance.

Floor surfaces with around 17–52 μm and 35–52 μm in $\boldsymbol{R_a}$ roughness parameter most likely represented the lower and upper bounds of operational ranges for the best slip resistance managements under the soapy and oily surface conditions, respectively. The test result also identified that the oily environment required twice rougher floor surface roughness than the soapy one in their lower boundary roughness scales: 35 μm versus 17 μm in $\boldsymbol{R_a}$ roughness parameter. On the other hand, the upper bound of floor surface roughness showed the same ranges of surface roughness scales: 52 μm in $\boldsymbol{R_a}$, 300 μm in $\boldsymbol{R_t}$, and 180 μm in $\boldsymbol{R_{tm}}$ roughness parameters under the three lubricated environments. However, there was a lack of correlations between the surface roughness and slip resistance properties under the clean and dry and wet surface conditions.

The inclusive results from this chapter also clearly display that the proposed concept on the operational ranges for the floor surface roughness may have practical design implications for floors and floor coverings to reduce slip and fall hazards.

References

AS/NZS 4663. (2004). *Slip resistance measurements of existing pedestrian surfaces*. Australian Standards AS/NZS 4663.

Bewoor, A. K., & Kulkarni, V. A. (2009). *Metrology & measurement*, Tata McGraw-Hill Education.

British Standards Institution (BSI). (1988). *Assessment of surface texture, Part 1. Methods and instrumentation*. BS 1134. London, UK: British Standards Institution.

Chaffin, D. B., Woldstad, J. C., & Trujillo, A. (1992). Floor/shoe slip resistance measurement. *Journal of American Industrial Hygiene Association, 53*(5), 283–289.

Chang, W. R. (2001). The effects of surface roughness and contaminant on the dynamic friction of porcelain tile. *Applied Ergonomics, 32*(2), 173–184.

Chang, W. R. (2002). The effects of surface roughness and contaminants on the dynamic friction between porcelain tile and vulcanized rubber. *Safety Science, 40*, 577–591.

Chang, W. R., Cotnam, J. P., & Matz, S. (2003). Field evaluation of two commonly used slip meters. *Applied Ergonomics, 34*(1), 51–60.

Chang, W. R., Huang, Y. H., Li, K. W., Filiaggi, A., & Courtney, T. K. (2008). Assessing slipperiness in fast-food restaurants in the USA using friction variation, friction level and perception rating. *Applied Ergonomics, 39*(3), 359–367.

Chang, W. R., Kim, I. J., Manning, D. P., & Bunterngchit, Y. (2001). The role of surface roughness in the measurement of slipperiness. *Ergonomics, 44*, 1200–1216.

Derler, S., Kausch, F., & Huber, R. (2008). Analysis of factors influencing the friction coefficients of shoe sole materials. *Safety Science, 46*(5), 822–832.

Harris, G. W., & Shaw, S. R. (1988). Slip resistance of floors: Users' opinions, Tortus instrument readings and roughness measurement. *Journal of Occupational Accidents, 9*(4), 287–298.

Hoang, K., & Stevenson, M. G. (1981). *Measurement of slipperiness of walkway surfaces*. Report. IE (University of New South Wales. School of Mechanical and Industrial Engineering); No. 1981. 1.

Hoang, K., Stevenson, M. G., Nhieu, J., & Bunterngchit, Y. (1987). *Dynamic friction at heel strike between a range of protective footwear and non-slip floor surface* (Report No. CSS/1/87). Center for Safety Science, University of New South Wales, Australia, July.

Hoang, K., Stevenson, M. G., & Willgoss, R. (1985). Measurement of dynamic friction between shoe soles and walkway surfaces. In *Proceedings of the 22nd Annual Conference of the Ergonomics Society of ANZ* (pp. 265–171). Toowoomba, December.

International Organization for Standardization (ISO). (1998). *Geometrical product specifications (GPS)—Surface texture: Profile method—Terms, definitions and surface texture parameters.* ISO 4287: 1997. Switzerland: International Organization of Standardization.

Johnson, K. L. (1985). *Contact mechanics*. Oxford, Clarendon: Cambridge University Press.

Kim, I. J. (1996a). Tribological concepts for the investigation of the pedestrian slipping and falling accidents—Part I. In *International occupational injury symposium* (p. 101). Sydney, Australia. February.

Kim, I. J. (1996b). Tribological approach for the analysis of pedestrian slip hazard—II. In *Proceedings of the '96 Spring conference of Korean Institute of industrial engineers*, April, Seoul, Korea, 279–285.

Kim, I. J. (1996c). Microscopic investigation to analyze the slip resistance of shoes. In *Proceedings of the 4th Pan Pacific conference on occupational ergonomics* (pp. 68–73). Taiwan, ROC, November.

Kim, I. J. (1996d). Microscopic observation of shoe heels for pedestrian slip hazard investigation. In *Proceedings of the 1st annual international conference on industrial application and practice* (pp. 243–250). Texas, USA, December.

Kim, I. J., & Smith R. (1999). The relationship between wear, surface topography characteristics and coefficient of friction as a means of assessing the slip hazards. In *2nd Asia-Pacific conference on industrial engineering and management systems* (*APIEMS'99*) (pp. 155–161). Ashikaga, Japan. October.

Kim, I. J. (2004a). Development of a new analyzing model for quantifying pedestrian slip resistance characteristics: Part I—Basic concepts and theories. *Industrial Journal of Industrial Ergonomics, 33*(5), 395–401.

Kim, I. J. (2004b). Development of a new analyzing model for quantifying pedestrian slip resistance characteristics: Part II—Experiments and validations. *Industrial Journal of Industrial Ergonomics, 33*(5), 403–414.

Kim, I. J. (2005a). Combined effects of friction and wear behaviors on fall safety measures. In *Contemporary ergonomics 2005* (pp. 498–502). Taylor & Francis.

Kim, I. J. (2005b). A new understanding on the shoe wear mechanism and its significance on slip resistance property. In *Contemporary ergonomics 2005* (503–508). Taylor & Francis.

Kim, I. J. (2006a). The current hiatus in fall safety measures. In W. Karwowski (Ed.), *International encyclopedia of ergonomics and human factors-2005* (pp. 2572–2576). USA: Taylor & Francis Group, LLC.
Kim, I. J. (2006b). A new paradigm for characterizing slip resistance properties. In W. Karwowski (Ed.), *International encyclopedia of ergonomics and human factors-2005* (pp. 2735–2740). USA: Taylor & Francis Group, LLC.
Kim, I. J. (2015a). Practical design search for optimal floor surface finishes to prevent fall incidents. In B. Evans (Ed.), *Accidental falls: Risk factors, prevention strategies and long-term outcomes* (pp. 80–103). Hauppauge, NY, USA: Nova Science Publishers, Inc. (Chapter 5).
Kim, I. J. (2015b). Wear observation of shoe surfaces: Application for slip and fall safety assessments. *Tribology Transactions, 58*(3), 407–417.
Kim, I. J. (2015b). Slip-resistance measurements for assessing pedestrian falls: Facts and fallacies. In B. Evans (Ed.), *Accidental falls: Risk factors, prevention strategies and long-term outcomes* (pp. 105–125). Hauppauge, NY, USA: Nova Science Publishers, Inc. (Chapter 6).
Kim, I. J. (2015d). Research challenges on slip-resistance measurements for assessing pedestrian fall incidents. *Journal of Ergonomics, 5*(3). doi:10.4172/2165-7556.1000e142
Kim, I. J. (2016a). A study on wear development of floor surfaces: Impact on pedestrian walkway slip-resistance performance. *Tribology International, 95,* 316–323.
Kim, I. J. (2016b). Identifying shoe wear mechanisms and associated tribological characteristics: The importance for slip resistance evaluation. *Wear, 360–361,* 77–86.
Kim, I. J., Hsiao, H., & Simeonov, P. (2013). Functional levels of floor surface roughness for the prevention of slips and falls: clean-and-dry and soapsuds-covered wet surfaces. *Applied Ergonomics, 44*(1), 58–64.
Kim, I. J., & Nagata, H. (2008a). Nature of the shoe wear: Its uniqueness, complexity and effects on slip resistance properties. In *Contemporary Ergonomics 2008* (Vol. 15, pp. 728–734). Taylor & Francis.
Kim, I. J., & Nagata, H. (2008b). Research on slip resistance measurements—A new challenge. *Industry Health, 46*(1), 66–76.
Kim, I. J., & Smith, R. (1998a). A study of the comparative geometry mating between the surfaces of the shoe and floor in pedestrian slip resistance measurements. In *The 5th Pan-Pacific conference on occupational ergonomics* (pp. 34–37). Kitakyushu, Japan. July.
Kim, I. J., & Smith, R. (1998b). Tribological characterization of the frictional force component in pedestrian slip resistance measurements. In *Third world congress of biomechanics* (*WCB'98*). Hokkaido, Japan. August.
Kim, I. J., & Smith, R. (2000). Observation of the floor surface topography changes in pedestrian slip resistance measurements. *Industrial Journal of Industrial Ergonomics, 26*(6), 581–601.
Kim, I. J., & Smith, R. (2001a). A critical analysis on the friction measuring concept for slip resistance evaluation. In *ASTM symposium on the metrology of pedestrian locomotion and slip resistance* (pp. 1–14). West Conshodocken, Pennsylvania, USA: ASTM Headquarters. June.
Kim, I. J., & Smith, R. (2001b). A study for characterising topography changes of shoe surfaces in the early stage of slip resistance measurements—Bearing area curve. In *6th Pan-Pacific conference on occupational ergonomics* (pp. 299–303). Beijing, P.R. China. August.
Kim, I. J., & Smith, R. (2001c). Three-dimensional analysis of floor surface wear during slip resistance measurements. In *6th Pan-Pacific conference on occupational ergonomics* (304–308). Beijing, P.R. China. August.
Kim, I. J., & Smith, R. (2003). A critical analysis of the relationship between shoe-floor wear and pedestrian/walkway slip resistance. In: Marpet, M. I., & Sapienza, M.A. (Eds.), *Metrology of pedestrian locomotion and slip resistance, American society of testing and materials.* Special Technical Publication (Vol. 1424, pp. 33–48). Philadelphia, USA: ASTM International.
Kim, I. J., Smith, R., & Nagata, H. (2001). Microscopic observations of the progressive wear on the shoe surfaces which affect the slip resistance characteristics. *Industrial Journal of Industrial Ergonomics, 28*(1), 17–29.
Leclercq, S., & Saulnier, H. (2002). Floor slip resistance changes in food sector workshops: prevailing role played by fouling. *Safety Science, 40*(7–8), 659–673.

Li, K. W., Chang, W. R., Leamon, T. B., & Chen, C. J. (2004). Floor slipperiness measurement: friction coefficient, roughness of floors, and subjective perception under spillage conditions. *Safety Science, 42*(6), 547–565.

Manning, D. P., & Jones, C. (1994). The superior slip resistance of footwear soling compound T66/103. *Safety Science, 18*(1), 45–60.

Manning, D. P., & Jones, C. (2001). The effect of roughness, floor polish, water, oil and ice on underfoot friction: Current safety footwear solings are less slip resistant than microcellular polyurethane. *Applied Ergonomics, 32*(2), 185–196.

Moore, D. F. (1972). *The friction and lubrication of elastomers, International series of monographs on materials science and technology* (Vol. 9). Headington Hill Hall, Oxford: Pergamon Press Ltd.

Moore, D. F. (1975). *Principles and application of tribology*. Oxford: Pergamon Press.

Perkins, P. J., & Wilson, M. P. (1983). Slip resistance testing of shoes—New development. *Ergonomics, 26*(1), 73–82.

Proctor, T. D. (1993). Slipping accidents in Great Britain—An update. *Safety Science, 16*(3–4), 367–377.

Stachowiak, G., & Batchelor, A. (2005). *Engineering tribology* (3rd ed., pp. 461–499). Butterworth-Heinemann, Elsevier Inc. (Chapter 10).

Strandberg, L. (1983). On accident analysis and slip-resistance measurement. *Ergonomics, 26*(1), 11–32.

Strandberg, L., & Lanshammar, H. (1981). The dynamics of slipping accidents. *Journal of Occupational Accidents, 3*, 153–162.

Thomas, T. R. (1999). *Rough surfaces* (2nd ed.). London, UK: Imperial College Press.

Chapter 8
Future Works

8.1 Introduction

This book introduces novel concepts on slip resistance measurements for the pedestrian fall safety assessments from a viewpoint of tribology. Whilst significant questions on the pedestrian fall safety evaluations still remain, this book proposes a foundation to understand complex characteristics and mechanisms of frictional and wear behaviours of the shoes and floors and their interactive effects on slip resistance performance amongst shoes, floors, and environments.

The inclusive contents from this book clearly demonstrate that measurements and interpretations of slip resistance properties amongst the shoe, floor, and environment should be based on thorough comprehension of the relevant mechanics and mechanisms involved at the sliding interfaces amongst the shoes, floors, and environments as an essential prerequisite. In this sense, it seems to be a valuable attempt to study their behaviours, characteristics, mechanisms, and related tribo-physical events of the shoes and floors and their overall impacts on slip resistance performance.

In order to understand tribo-physical processes involved at the sliding interface between the shoe and the floor surface, safety researchers and practitioners must recognise how two pairing surfaces interrelate when they are loaded together. Thus, surface structures and relevant contextual information on surface interactions were comprehensively reviewed and discussed with measuring instruments to quantify topographic characteristics of the shoes and floors in this book.

Based on the extensive overviews on frictional and wear behaviours of two dissimilar materials, their accommodating mechanisms, and related theories and concepts, several theoretical assumptions and geometrical models were suggested and developed for the shoe-floor sliding friction system. A solid theoretic model to predict slip resistance performance amongst the shoes, floors, and environments has not developed yet, but this book proposes that tribological approaches seem to be a

I.-J. Kim, *Pedestrian Fall Safety Assessments*,
DOI 10.1007/978-3-319-56242-1_8

worthwhile attempt to overcome limitations on the current researches and industry practices for the measurements of slip resistance properties.

It also needs to be mentioned that this book clearly showed the effects of floor surface finishes on slip resistance properties and identified operational ranges of floor surface roughness for optimal slip resistance performance under different slippery walking environments. The suggested concepts on operational ranges with lower and upper bounds for the floor surface roughness may have practical design implications for the floors and floor coverings to reduce slip and fall hazards.

It is considered that collected information on operative ranges of floor surface roughness under diverse walking environments would be beneficial to develop specific design information for floor surface finishes required to preventing pedestrian slip and fall incidents under a range of walking environments and circumstances.

Therefore, thorough understandings on the slip resistance forces generated between the footwear and underfoot surfaces with their friction and wear behaviours would be worthwhile challenges to improve the current design practices and design related issues such as guidelines and policies for the floors/walkways.

Finally, it is wished that this book may provide an insight to fill the current knowledge gaps and misinterpretations on slip resistance measurements that are mainly caused by oversimplified perceptions on tribo-physical phenomena amongst the shoes, floors, and environments.

8.2 Review of Overall Aims

The overall purposes of this book were:

(1) to identify major problems of the existing concepts, theories, and methodologies for the evaluation of slip safety amongst the shoe heels, floor surfaces, and environments;
(2) to characterise and analyse slip resistance properties between the shoe and floor surface from a tribo-physical point of view.
(3) to understand friction and wear behaviours involved at the sliding interface between the shoe and floor surface;
(4) to analyse surface topographies of the shoes and floors and assess their variations with sliding friction events and their effects on the slip resistance properties;
(5) to investigate geometric interactions between the shoe heel and floor surface during the repetitive sliding process;
(6) to suggest novel concepts and theory models characterise slip resistance properties, which are more objective and reliable than a simple friction measurement; and

(7) to verify new concepts, theory models and methods to predict ways in which the slip resistance properties may be measurable more proficiently and accurately.

To achieve the above goals, comprehensive investigations of the surface topographies of both shoes and floors, analyses of mechanical and physical characteristics of both bodies, explorations of related tribo-physical phenomena, and microscopic works were carried out over the seven chapters in this book.

8.3 Recommendations for the Future Studies

Whilst this book successfully undertook its primary purposes of improving our understanding of slip resistance properties for the fall safety assessments, additional works are also required to ensure that these efforts eventually result in substantial reductions of injuries and fatalities due to slip and fall incidents.

Specifically, it can be considered that physical accuracy and validity of the suggested shoe-floor-environment sliding friction model need to be improved and expanded to include a diverse range of shoes, floor types, and environmental conditions. The following sub-sections propose specific ideas on how the future generation of the shoe-floor-environment tribo-physical model(s) can be further improved to become a useful tool(s) for the prevention of slip and fall incidence.

8.3.1 Necessary Advancements in the Tribo-physical Model

The suggested shoe-floor tribo-physical model in this book needs to be improved to becoming a widely-accepted tool for the pedestrian fall safety assessments. The enhancement demands to increase physical accuracies of the model, particularly by including elastohydrodynamic lubrication (EHL) effects at the sliding interface. In addition, the model also needs to be expanded to demonstrate wear behaviours of the shoes and floors and their interactive effects on slip resistance performance.

However, one of the major limitations for the development of the shoe-floor friction model as described in this book was that deformation effects due to hydrodynamic pressures were ignored. Therefore, the EHL effects which seem to play a critical role in the shoe-floor-contaminant interface need to be included in the future modelling efforts.

Including the EHL effects in the shoe-floor-environment friction model must consider the issue of film thicknesses generated at the contact-sliding interface as a function of fluid pressures. Whilst fluid pressures across the shoe surface can be approached using the Reynolds equation as a system of linear equations, solving the equation requires highly non-linear equations. Thus, introducing EHL influences into the shoe-floor-pollutant friction model requires additional research to identify an algorithm that is capable of solving this set of non-linear equations.

Significant challenges also exist to develop the shoe-floor-contaminant tribo-physical model. To ensure that the tribo-physical model accounts for transient effects, the Reynolds equation has to be modified to include squeeze film effects. As the complexity of the friction model increases, new testing protocols also need to be developed to validate both subsequent models. For example, an eventual shoe-floor-environment friction model to capture transient effects will need to be validated with experiments that can capture the transient effects. In addition, a new slip testing device would have to be developed to better capture transient effects. Thus, significant questions still remain to improve the currently developed shoe-floor-contaminant model.

8.3.2 *Long Term Plan for the Tribo-physical Model*

Although the long-term goal for this research effort would be beyond this book, it definitely needs to develop a predictive tool that can prevent slip and fall accidents. Development of an integrated shoe-floor-contaminant tribo-physical model that can capture the effects of multifactorial characteristics such as macro-geometries (tread patterns), shoe and floor surface finishes, fluid properties, and various loading conditions (shoe-floor contact angle, normal force, heel velocity) seems to be an ideal for achieving this goal. The envisioned shoe-floor-contaminant tribo-physical model will become an enhanced method to:

(1) improve our understanding of how multifaceted factors affect slip resistance performance amongst the shoes, floors, and contaminants.
(2) deliver improved slip-resistance data for a range of shoes, floors, and environmental combinations.
(3) provide reliable and valid information on slip resistance properties amongst the shoes, floors and environments to the safety researchers, practitioners, and consultants.
(4) aid in improving designs for the shoe soles/heels and floor surface finishes that require different levels of surface roughness for different types of environmental conditions to effectively control slippery walking environments.

8.3.3 *Improvement for Slip Measuring Concepts*

The current slip measuring devices and instruments have numerous limitations, particularly due to their incapability to accurately simulate slipping and falling conditions and to systematically account for the development of friction and wear behaviours of the shoe and floor surfaces. Initial surface features and topographic characteristics of both shoes and floors are frequently and significantly modified by repetitive friction and wear developments. As a result, slip resistance properties become noisy and continuously change as a function of a complex array of tribo-physical phenomena amongst the shoes, floors, and environments. Thus, a single index of friction measurement as a form of COF does not provide a reliable determination of essential properties for slip resistance amongst the shoes, floors, and environments and accordingly has obvious difficulty as an indicator for the fall safety measures.

The present approach for measuring slip resistance performance simply implies that a larger friction coefficient indicates better slip resistance, yet the amount of required friction even whilst walking on a severely sloped incline rarely exceeds 0.4. Therefore, designs for the shoe soles/heels and floor surfaces that result in COFs much above 0.4 may not improve slip resistance as would be indicated by the traditional method of measuring slip resistance.

In fact, excessive levels of slip resistance may impede safe and comfortable ambulation during the swing portion of gait and cause a trip (Chaffin et al. 1992). An improved attempt to measuring slip resistance properties need to evaluate how the shoe and floor surfaces perform in the presence of different pollutants under a range of loading circumstances, which seem to be more representative of the variable conditions experienced by both shoe and floor surfaces.

Development of a computational model for the shoe-floor-environment friction requires that COFs should efficiently measure a large range of conditions, which may provide a complete picture of slip resistance properties for the shoe-floor combination. Therefore, once the shoe-floor-contaminant tribo-physical model is developed, different tests need to be created so that this integrated model can evaluate the slip resistant properties across a wide range of conditions.

The tribo-physical model also needs to be a valuable tool to the shoe and floor manufacturers for designing shoe heels/soles and floor surfaces with improved slip resistance properties. The shoe and floor manufacturers and industries will benefit to evaluate slip resistance performance efficiently and effectively across a large range of shoe-floor-environment conditions relevant to slip and fall accidents.

In addition, shoe manufacturers need to rapidly test their shoe designs against many different floor surfaces and contaminants to determine their shoe functions to maintain adequate slip resistance against a range of different walking conditions. Similarly, flooring design can also be tested rapidly against many pollutants and shoe types. A computational shoe-floor-contaminant friction model also should help shoe designers to identify the heel/sole areas of shoes with the highest hydrodynamic pressures. This feature will allow the shoe manufacturers to focus on redesigning the tread patterns in these regions to relieve the peak hydrodynamic pressures.

A fully developed shoe-floor-contaminant tribo-physical model may also be useful to evaluate how biomechanical factors affect the tribological characteristics such as friction and wear mechanisms of shoes and floors. Therefore, future research urgently requires to developing a refined computational model for the shoe-floor-contaminant friction system to assess different walking styles and strategies to determine which is capable of maximising slip resistance performance against a range of shoe-floor-environment combinations.

8.4 Conclusions

This book suggests various theoretical concepts and models analyse comprehensive tribological characteristics of the sliding interfaces amongst diverse ranges of shoes, floors, and environmental conditions. Surface analyses with derived numerical quantifications clearly reveal that the slip resistance properties are affected by multiple factors and their geometrical interactions are too complex to understand. Without in-depth considerations on the tribo-physical characteristics amongst the shoe, floor, and environment, therefore, a simple format of friction measurement is not a reliable way anymore to evaluate slip resistance properties and consequently slip hazards.

This book was focused on broadening the knowledge base and developing novel concepts on which improvements in the validity and reliability of the slip resistance measurement might be made. To achieve this goal, the problem was critically assessed and approached from a tribological point of view where a principal understanding of the shoe-floor friction and wear mechanisms could be made. The main concepts included a comprehensive investigation of surface topographies of both shoes and floor surfaces and their interactive effects on slip resistance performance. This book suggested new theoretical concepts and models for the shoe-floor-contaminant tribo-physical behaviours to analysing the slip resistance properties.

This book also suggested a model on operational ranges with lower and upper bounds for the floor surface roughness. Collected information on operative ranges of floor surface roughness under diverse walking environments would be beneficial to develop practical design information for the floor surface finishes required to preventing pedestrian slip and fall incidents.

It is wished that this book may provide a new insight to fill the current knowledge gaps on slip resistance measurements that are mainly caused by oversimplified views on the complex nature of friction phenomena at the sliding interface amongst the shoe, floor, and environment and their related tribo-physical effects on slip resistance performance.

Reference

Chaffin, D. B., Woldstad, J. C., & Trujillo, A. (1992). Floor/shoe slip resistance measurement. *Journal of American Industrial Hygiene Association, 53*(5), 283–289.

Index

I.-J. Kim, *Pedestrian Fall Safety Assessments*,
DOI 10.1007/978-3-319-56242-1

L

M

Q

MIX
Papier aus verantwortungsvollen Quellen
Paper from responsible sources
FSC® C105338

If you have any concerns about our products,
you can contact us on
ProductSafety@springernature.com

In case Publisher is established outside the EU,
the EU authorized representative is:
Springer Nature Customer Service Center GmbH
Europaplatz 3, 69115 Heidelberg, Germany

Printed by Libri Plureos GmbH
in Hamburg, Germany